The Senses of Modernism

ALSO BY SARA DANIUS

Prousts Motor

The Senses of Modernism

Technology, Perception, and Aesthetics

Sara Danius

Cornell University Press   *Ithaca & London*

First published 2002 by Cornell University Press
First printing, Cornell Paperbacks, 2002

Printed in the United States of America

Library of Congress Cataloging-in-Publication Data

Danius, Sara.
    The senses of modernism : technology, perception, and aesthetics/
Sara Danius.
        p. cm.
    ISBN 0-8014-3899-3 (cloth : alk. paper)
    ISBN 0-8014-8800-1 (pbk. : alk. paper)
    1. Modernism (Literature) 2. Modernism (Aesthetics) I. Title.
    PN56 .M54 D36 2002
    809' .9112—DC21

                                            2001007787

Cornell University Press strives to use environmentally responsible suppliers and materials to the fullest extent possible in the publishing of its books. Such materials include vegetable-based, low-VOC inks and acid-free papers that are recycled, totally chlorine-free, or partly composed of nonwood fibers. For further information, visit our website at www.cornellpress.cornell.edu.

Cloth printing   10 9 8 7 6 5 4 3 2 1
Paperback printing   10 9 8 7 6 5 4 3 2 1

In memory of my father,

Lars Danius

(1907–1996)

Contents

Acknowledgments

In the course of writing this book I have incurred many debts, intellectual and otherwise. I would like to extend a special thanks to Fredric Jameson, who inspired this project in numerous ways. He was a singular teacher and extraordinary reader, and I am grateful for his intellectual example. Without his commitment to my project, this book would not have been possible. To Bengt Landgren, a model of savoir-faire and an exceptionally generous critic, I also owe a special debt of gratitude. His vast knowledge of modernism, expertise in literary theory, and shrewd observations have enriched my work. Barbara Herrnstein Smith introduced me to new ways of thinking about the art and theories of technoscience. An exemplary reader, she scrutinized earlier versions of this book, always hitting the mark. I have learnt a great deal from her about how to craft an argument. I am equally grateful to Toril Moi, for her friendship, exacting readings, and hard questions. She supported my work at every step of the way, from the initial stirrings to finished manuscript. She is a true source of intellectual inspiration. I also want to thank Valentin Y. Mudimbe, Michael Hardt, Torsten Pettersson, and the late Naomi Schor for their gracious assistance and valuable suggestions.

I have benefited from the insight of friends and colleagues, sometimes without them knowing it. My great friends Anders Stephanson and Rebecca Karl read through early versions of the manuscript, and I have learned much from their extensive observations. A series of conversations about cinema studies proved essential: my thanks to Catherine Benamou, Norman Klein, Jennifer Parchesky, and Tom Whiteside. I have also benefited from exchanges with Keith Wailoo about the history of medical technology. For various kinds of assistance, I thank Åsa Beckman, Jonathan Beller, Renu Bora, Marta Braun, Nina Danius, Hans-Göran Ekman, Aris Fioretos, Stefan Helgesson, Julia Hell, Tina Jonsson, Eleanor Kaufman, Arne Melberg, Reinhart Meyer-Kalkus, Hank Okazaki, Hans Ruin, Charles Salas, Yaël Schlick, Per Stam, Neferti Tadiar, Astrid Söderbergh Widding, and Xudong Zhang. I would also like to thank Catherine Rice, Teresa Jesionowski, Melissa Oravec, and the reviewers at Cornell University Press for their commitment to my project.

I am grateful to the Getty Research Institute, and especially to Michael Roth, for their support when I was completing the manuscript and researched modernist art and other visual material for my book. The staff of the Department of Photographs at the Getty Museum was equally graceful. My sincere thanks also to the Wissenschaftskolleg zu Berlin.

I have had the privilege of working in a series of extraordinary libraries and have frequented their interlibrary loan desks in particular. I wish to thank the excellent staff of Perkins and Lilly Libraries at Duke University in Durham, North Carolina; Carolina Rediviva at Uppsala University and the Royal Library in Stockholm, Sweden; the Research Library, including the Special Collections, at the Getty Center in Los Angeles, California; and at the Library at Wissenschaftskolleg in Berlin, Germany.

My research has been supported by grants from the Fulbright Commission for Educational Exchange; Knut and Alice Wallenberg Foundation; American-Scandinavian Foundation, including Thord-Gray Memorial Fund and Håkan Björnström Steffanson Fund; Swedish Institute; Jacob Letterstedt Fund; Lars Hierta Memorial Fund; Karl and Betty Warburg Fund; and the Swedish Council for Research in the Humanities and Social Sciences. I have also received a Mellon Dissertation-Year Fellowship from Duke University Graduate School. I acknowledge these sources with gratitude.

My greatest debts are to Stefan Jonsson, my husband and colleague, for years of insightful, challenging, and pleasurable conversation. He has read and commented on every single draft, and contributed to my work in invaluable ways. I also want to thank my mother, Anna Wahlgren. Her stamina and fearlessness as a writer are exemplary.

This book is dedicated to the memory of my father, Lars Danius, who did not live long enough to see its completion. Born in 1907, his life spanned the twentieth century, and he devoted it to the study of history, language, and literature. His love always sustained me, and always will.

Earlier versions of several chapters have appeared in the following publications: *boundary* 2, 27, no. 2 (summer 2000); *Forum for Modern Language Studies* 37, no. 2 (2001); *Modernism/modernity* 8, no. 2 (2001); *Proust in Perspective: Visions and Revisions*, ed. Armine Kotin Mortimer and Katherine Kolb (Urbana: University of Illinois Press, 2002). I am grateful for permission to reprint.

S.D.

Berlin
December 2001

Abbreviations

Z   Thomas Mann. *Der Zauberberg*. Frankfurt am Main: Fischer, 1991.

MM   Thomas Mann. *The Magic Mountain*. Translated by John E. Woods. New York: A. Knopf, 1995.

RTP   Marcel Proust. *A la recherche du temps perdu*. 4 vols. Edited by Jean-Yves Tadié et al. Bibliothèque de la Pléiade. Paris: Gallimard, 1987–1989.

REM   Marcel Proust. *Remembrance of Things Past*. 3 vols. Translated by C. K. Scott Moncrieff and Terence Kilmartin. New York: Vintage, 1982.

CSB   Marcel Proust. *Contre Sainte-Beuve, précédé de Pastiches et mélanges et suivi de Essais et articles*. Edited by Pierre Clarac and Yves Sandre. Bibliothèque de la Pléiade. Paris: Gallimard, 1977.

U   James Joyce. *Ulysses*. Edited by Hans Walter Gabler et al. New York: Random House, 1986.

Unless otherwise noted, translations are my own.

The Senses of Modernism

Introduction

Orpheus and the Machine

Friedrich Nietzsche once contemplated what he called the premises of the machine age. "The press, the machine, the railway, the telegraph are premises whose thousand-year conclusion no one has yet dared to draw," he wrote in 1878.[1] Nietzsche's words are as valid as ever, and in the age of cybernetics, they raise the question once again: What are the relations of modernization and modes of experience? More specifically, what are the relations of technological change and aesthetics, and how may they be conceptualized?

This is a book about technology and aesthetic modernism. It takes a closer new look at high modernism, opening a hitherto unexplored domain in the study of modernism and modernity: the nexus of perception, technological change, and literary form. Moving within a historical trajectory that extends from, roughly, 1880 to 1930, I argue that modernist aesthetics from Marcel Proust to James Joyce is an index of a technologically mediated crisis of the senses, a perceptual crisis that ultimately cuts across the question of art as such.

I began this book with a hunch. My original intention was to explore how, in the modernist period, the human sensorium came to be invoked as a touchstone for aesthetic gratification and experiential authenticity. In the process, it became clear that the machine invariably appeared in the same thematic cluster as the corporeal, the sensory, and the aesthetic, a complex of problems which offered an infinitely more interesting object of study. I decided to explore the thesis that the specific aesthetics of perception on which so much of classical modernism turns is tightly bound up with modern machine culture, and nowhere so strikingly as in those contexts where the technological has commonly been seen as irrelevant or

1

as having no aesthetic raison d'être—Proust's phenomenology of perception being a case in point.

What became apparent was that classical modernism represents a shift from idealist theories of aesthetic experience to materialist ones, or, which ultimately amounts to the same thing, that the emergence of modernist aesthetics signifies the increasing internalization of technological matrices of perception. For to chart how the question of perception, notably sight and hearing, is configured in the modernist period is to witness the ever-closer relationship between the sensuous and the technological. This book, then, chronicles the adventures of the eye and the ear in the early twentieth century. Tracing a crucial moment in the history of aesthetics, it maps the specific ways in which aesthetic developments intersect with technological developments. Moving from Ruskin to Vertov, from Röntgen to Proust, my discussion ultimately suggests that modernist culture prefigures today's virtual realities of perception.

The modernist era coincides with the historical period that saw the emergence of, among other things, phonography, chronophotography, cinematography, radiography, telephony, electricity, and technologies of speed.[2] Yet traditional accounts of modernism, marked as they often have been by an antitechnological bias, have obscured this coexistence. To be sure, most scholars and critics agree that avant-garde movements such as cubism, futurism, surrealism, and vorticism must be understood in relation to technology.[3] Similarly, it is generally agreed that the emergence of mass culture, notably cinematography and recorded music, must be grasped against the background of the second industrial revolution. As for the novel, however, the notion of aesthetic autonomy still looms large on the interpretive horizon, preempting, it seems, reflection on the historicity of high-modernist aesthetic practices. This makes the high-modernist novel, along with poetry, an exceptionally appropriate field for the study of the relations of technology and aesthetics, since it is precisely in this domain that the incompatibility between the two has been most strongly asserted.

Accordingly, one of the aims of this book is to challenge the widespread assumption that there is a gap between a merely technological culture and what has been seen as a more properly aesthetic one. The second machine age and its technoscientific discourses, particularly those pertaining to technologies of perception, have to be reckoned into a historical understanding of high-modernist aesthetics, and I take such an approach to yield a richer and more accurate understanding of modernism. The argument brought forward here is not that technology provides a crucial *context*, although it does that too, but a more radical one: that technology is in a specific sense *constitutive* of high-modernist aesthetics. A first task, then, is to critique and dissolve a standard dualism in accounts of modernism.

A second and even more important task is to develop a theoretical framework that allows a commonly overlooked complex of problems to come into focus: how high-modernist aesthetics is inseparable from a historically specific crisis of the senses, a sensory crisis sparked by, among other things, late-nineteenth- and early-twentieth-century technological innovations, particularly technologies of perception. In tracing this complex of problems across so many exemplary texts, I want to argue that this crisis is a fundamental marker of the high-modernist moment and some of its most vital aesthetic innovations. The basic propositions of my inquiry thus revolve around questions of sensory perception or, to be more precise, around the specific ways in which categories of perceiving and knowing are reconfigured in a historical situation in which technological devices are capable of storing, transmitting, and reproducing sense data, at the same time articulating new perceptual and epistemic realms.[4] Hence the problem that so many modernist texts and artifacts stubbornly engage: how to represent authentic experience in an age in which the category of experience itself has become a problem.

My primary exhibits are three works of high modernism, Thomas Mann's *The Magic Mountain* (1924), Marcel Proust's *Remembrance of Things Past* (1913–27), and James Joyce's *Ulysses* (1922). I have selected these novels because they are extraordinarily rich and challenging. I have also chosen them because of their canonical status. In traditional literary history, these texts are generally thought of as defining the quintessence of the modernist novel. The names of Mann, Proust, and Joyce, like those of Woolf, Kafka, Musil, Broch, Faulkner, and dos Passos, carry a certain representational force, and for this reason my selection is a strategic one as well. What is more, I have chosen these narratives because they describe a general transition from technological *prosthesis* to technological *aisthesis*, thus moving from externalization to internalization.

I also want to bring together these modernist monuments because they project themselves as literary equivalents of the Wagnerian *Gesamtkunstwerk*, as synaesthetic works of art seeking to transcend genre.[5] I thus take these narratives to be characteristic of a particular strand of modernism, that of the Book of the World.[6] This means that the tales of Mann, Proust, and Joyce articulate an aesthetic program for which the text itself serves as an example. Moreover, in advertising themselves as total works of art, they indirectly present themselves as literary artifacts that may appeal to all the senses. In the age of sensory dissociation and reification, such a synaesthetic ideal is no coincidence. This essentially formal aspect alone motivates a historicizing approach. Indeed, the sheer attempt to turn the novel into an encyclopedic, totalizing, all-inclusive work of art is a symptom of the profound crisis that the genre is facing in this period. These novels thus broach a topic that grants them a pivotal role in any inquiry into the relations of technology and modernist aesthetics, namely, the status of the novel and of art as such. And since Mann, Proust, and Joyce can be said to narrate the seismographic changes that affect culture, high and low, in the age of technological reproducibility, my textual analyses approach their novels as allegories of their own historical conditions of possibility. Such a perspective could easily be extended to numerous other modernist writers, in particular Robert Musil, Hermann Broch, and John dos Passos, but also to writers in whose works the totalizing impulse is less pronounced, such as Virginia Woolf, Djuna Barnes, Joseph Conrad, Wyndham Lewis, and Louis Aragon.

I should stress that my intention is neither to catalogue individual authors' attitudes toward the second machine age nor to provide an inventory of technological motifs in modernist literature. The analyses that follow are also not designed to merely complement "internalist" with "externalist" perspectives. The burden of the present book is to demonstrate that in the final analysis, the conventional distinction between the external and internal in studies of literary modernism derives from an aesthetic division of labor that is itself preeminently modernist.

Modernism and Modernity

In recent years, theorists and critics have begun the immense project of revising the very terms in which the historiography of modernist art and culture has typically been embedded.[7] Several studies also chart how modernist practices reflect technological change, thereby addressing some of the problems I explore in this book.[8] My choice of texts aside, *The Senses of Modernism* is the first to delineate a dialectical understanding of high-modernist aesthetics and technologies of perception. Centrally concerned with aesthetics and questions of form, I seek to demonstrate, as Italo Calvino once wrote of *Remembrance of Things Past*, that the advent of modern technology "is not just part of the 'color of the times,' but part of the work's very form, of its inner logic, of the author's anxiety to plumb the multiplicity of the writable within the briefness of life that consumes it."[9] In focusing on the relations of technological change, modes of representing the human sensory apparatus, and the fabric of modernist texts, *The Senses of Modernism* is thus an inquiry into the historical mediation of form.

My book combines a historicizing approach with close textual analysis. I believe that to be fully convincing, and to avoid reductionism, any serious attempt to investigate the relations of technology and literary modernism—in fact, any attempt to historicize modernism—has to be grounded in close textual analysis. From a methodological viewpoint, then, something like a stereoscopic device is needed, a dialectical analytics of the relations of modernism and modernity. In this spirit, Fredric Jameson has suggested that a rethinking of modernism has to be based on a comparative sociology of modernism and its cultures, underscoring that one has to work "both sides of the street and [dig one's] tunnel from both directions; one must, in other words, not only deduce modernism from modernization, but also scan the sedimented traces of modernization within the aesthetic work itself."[10] This is why it is imperative to pay close attention to the fine grain of the narrative text. This is also why I analyze a small number of literary texts in depth. It is one thing to limit the commentary to a set of motifs, images, or themes in modernism at large; it is quite another to take into account the relative autonomy of the narrative text as a whole.

Furthermore, *The Senses of Modernism* is predominantly concerned with technologies of perception. Photography, chronophotography, cinematography, and radiography are essential in my interrogation of high-modernist aesthetics, as are phonography and telephony. Technologies of speed such as the automobile are also major protagonists; I conceive of them as visual technologies or, to be more specific, as visual framing devices on wheels.[11] Although each of these technologies has its own relatively distinct history, they have a feature in common which grants them a central role in this book: they address, involve, or interfere with the sensory apparatus in more immediate ways than do, for example, production technologies, and ultimately raise questions having to do with truth, knowledge, and verification.[12] For this reason, technologies of perception also intersect with the question of aesthetics.

Modern theories of the aesthetic, Terry Eagleton reminds us, begin as a "discourse of the body." As originally conceived by the eighteenth-century philosopher Alexander Baumgarten, aesthetics is primarily concerned with the sensory infrastructure of the human body and only secondarily with the essence of art.[13] It is in the nature of aesthetics to be concerned with the ways in which the human body, through the senses, may derive gratification from certain artifacts and activities that belong to the phenomenal order. The etymological meaning of "aesthetics" springs out of a cluster of Greek words which designate activities of sensory perception in both a strictly physiological sense, as in "sensation," and a mental sense, as in "apprehension." *Aisthetikos* derives from *aistheta,* things perceptible by the senses, from *aisthesthai,* to perceive.[14] This is why I bring technologies of perception into focus.

Neither technology nor modernism are stable conceptual constructs. Technology is a name that has been assigned to various sets of phenomena, practices, and processes in various historical formations, be it productive forces, skills, tools, or know-how, to mention just a few possible denotations. The meanings of the phenomena named vary in different historical periods in Western society, from Aristotle's notion of *techne* as skill to the contemporary notion of technology as machinery and technics. In this book, "technology" is equivalent to "that which is called 'technology' in the discourses under discussion."[15]

Modernism, too, is a word with a notoriously vague meaning. In addition, there are significant national differences between the various conceptions of the nature of modernism and the avant-garde movements, including the histories of these conceptions.[16] A few heuristic definitions and delimitations are therefore in order here. This is primarily a study of European modernism, and by "modernism" I understand an umbrella term covering a variety of cultural practices, not merely experimental prose and poetry but also the avant-garde movements, including those involved in art, music, and the transformation of daily life.

For historical reasons, and not so much for stylistic or formal ones, I take the classic period of literary modernism to extend, roughly, from 1880 to 1930. When I refer to high modernism, or the high-modernist novel, which is the focal point of the critical-interpretive part of the book, I refer primarily to the 1920s. It was a remarkable decade, not only in the realm of literature. In 1922, Joyce published *Ulysses,* T. S. Eliot *The Waste Land,* Woolf *Jacob's Room,* Katherine Mansfield *The Garden Party,* Bertolt Brecht *Baal,* Proust *Cities of the Plain,* to mention just a few.[17] In the following year Rainer Maria Rilke published *Sonnets to Orpheus* and *Duino Elegies.* In 1924, Mann's *Magic Mountain* appeared, as did Breton's surrealist manifesto and E. M. Forster's *A Passage to India.* In 1925, Woolf's *Mrs Dalloway* and Franz Kafka's *The Process* were published.

The 1920s were also the decade when, for example, Edgard Varèse's noisy symphony *Amériques* was first performed, Man Ray invented his shadowy rayographs, and László Moholy-Nagy his photograms, Fernand Léger screened his visually advanced film *Ballet mécanique,* Dziga Vertov shot his self-referential movie *The Man with a Movie Camera,* and Sergei Eisenstein made *Potemkin.* Moreover, in the 1920s technological inventions became aesthetically gratifying objects in themselves, as witness the new objectivity movement in Germany in general and Bauhaus school photography in particular (figure 1). Last but not least, the decade saw the emergence of sound film.[18]

I take modernism to be already embedded in processes of modernization. That which goes under the name of modernism is here conceived of as so many forms of crisis management, not exterior to but already inscribed in those historical developments they sought to go beyond or else call into question. Accordingly, in advancing the hypothesis that a certain logic of technologization is inherent in high-modernist aesthetics, this book seeks to show that the innovations of modernism are situated in the same field of socioeconomic processes and technoscientific transformations that made mass culture possible.[19] For the antitechnological bias operates in conjunction with the clichéd notion that high-cultural art is inherently better and more deserving of scholarly attention because, among other reasons, it is less or not at all informed by technologized production, and likewise, that its low-cultural counterpart, mass culture, is inauthentic or mediocre because, among other reasons, it is far removed from the activity of the creative mind of the artist. This notion has informed a good number of the various theories of literary modernism.

The critical survey of theories of modernism that I offer in Chapter 1 is designed to show that, as a consequence of the insistence on aesthetic autonomy, the emergence of modernism has commonly been studied in isolation from modernity, even in those cases where modernity is posited as a "context" or "background." I call this conceptual pattern the topos of the split. It shares a family likeness with what Andreas Huyssen, in his work on modernism, specifically the avant-garde, has designated as the Great Divide, a discursive construct that builds on "the categorical distinction between high art and mass culture."[20] This distinction is intimately connected not only to the antitechnological bias of the influential paradigms of modernism that have articulated the nature and vocation of modernist aesthetics but also to a whole series of recurring dichotomies such as the opposition of art to society, of beauty to utility, of the organic to the me-

1. Marianne Brandt, [*Phonograph with Palm*], ca. 1926. J. Paul Getty Museum, Los Angeles. © 2001 Artists Rights Society (ARS), New York / VG Bild-Kunst, Bonn.

chanical, of aesthetic discourse to communicative discourse. If, then, high-modernist aesthetics is not exempt from a certain logic of technologization, and if such a logic is frequently associated with mass-cultural artifacts, then the question arises: How does this logic make itself felt in modernist writings, and, how should it be accounted for theoretically?

This book aims to develop a critical-interpretive framework that goes beyond those forms of causality that have dominated traditional descriptions of how modernism relates to modernity. As I argue in Chapter 1, a deterministic definition of the nature and operations of technology can be shown to inform the founding myth of modernism. Because of such determinism, technology inevitably figures as the *other* of modernist aesthetics. Whether modernism is conceived of simply as *determined by* modernity or is seen as a simple *reaction to* modernity, the relation of modernism to modernity, or aesthetics to technology, is posited as one of externality.[21] Such a construct, I believe, yields historically inaccurate views of the modernist moment. It is a mistake to assume, for example, that the interest in subjective time that makes itself felt in Bergson, Proust, Woolf, and other thinkers and writers in the modernist period is to be theorized as mere *reaction* against public or homogeneous time. There are at least two reasons for this.

First, the mobilization of private time may be seen as a reaction only insofar as it is recognized that such a reaction will inevitably bear the traces of that against which it reacts. In that case, however, the various modernist reactions have to be rewritten as the essential *invention* of a new set of terms or a new matrix of meanings that attaches itself to a recognized set of terms—for example, private and public, subjective and objective—which at the same time constitute the conceptual arena on which the reaction itself is played out.

Consider Junichiro Tanizaki's *In Praise of Shadows* (1933), a meditation on what the author thinks of as an essentially Japanese sensibility. Picking insignificant objects of contemplation such as women's complexions, toilets, lamps, paperware, lacquerware, and miso soup, Tanizaki stages a drama featuring two major agonistic characters—light versus shadow, or artificial light versus natural darkness. These then reveal themselves as allegorical vehicles for the conflict between traditional society and modern progress, that is, Japan and the West. Yet Tanizaki's reflection on traditional Japanese sensibility is, one might say, an invention of this same sensibility. No episode is more telling than when the narrator relates a visit to a restaurant in Kyoto whose dining rooms used to be lit by candlelight, now exchanged for electrical lamps. The narrator asks the waiter to bring a candle to the table, and so it is that he discovers, at one stroke, something he had not perceived before:

And I realized then that only in dim half-light is the true beauty of Japanese lacquerware revealed. The rooms at the Waranjiya are about nine feet square, the size of a comfortable little tearoom, and the alcove pillars and ceilings glow with a faint smoky luster, dark even in the light of the lamp. But in the still dimmer light of the candlestand, as I gazed at the trays and bowls standing in the shadows cast by that flickering point of flame, I discovered in the gloss of this lacquerware a depth and richness like that of a still, dark pond, a beauty I had not before seen.[22]

Such a passage, resounding with a Proustian pitch, makes clear how the advent of electricity has split open a new opposition between darkness and light, depth and surface, opacity and luster. Just as the introduction of electricity makes visible a vanished world that henceforth appears as suffused with darkness and soft shadows, so the intrusion of Western technologies helps fabricate a Japanese notion of the imagined community that henceforth appears as original, genuine, or traditional—when it has already been lost. A different matrix of perceptual possibilities has been put in place, hence also a new signifying system, and it is mediated by what Tanizaki designates as the Western ideology of progress. In a similar vein, one may argue that the nonrational, anti-instrumentalist, and subjectivist impulses dwelling within the various modernisms are intertwined with so-called rationalizing features, from Proust's elevation of sensory experience to Conrad's imperative to make the reader *see*, from Woolf's celebration of deep temporality to Joyce's pursuit of profane epiphanies.

Second, to maintain that the modernist explorations of subjective time are to be theorized as mere vested responses to the growing hegemony of homogeneous time is to reinvent one of the founding myths of high modernism: the disjunction between art and society, culture and history. And, more generally, even if it is true that high-modernist practices often sought to transcend or even cancel what was thought of as public time, instrumental reason, and the logic of commodification, it can nevertheless be demonstrated that those art forms were affected, enabled, and to some extent even created by those same developments.

The following pages, then, seek to articulate an alternative model for the relations of modernism and technology, one that allows for a notion of mutual determination or, alternatively, multiple determination. I propose to conceive of technology in terms of conditions of possibility, and a few cautionary words are in order here. By referring to this notion, I do not posit it in the transcendental sense, that is, as an ensemble of necessary-and-sufficient causes, but rather as a set of significant factors of contribution among many others. Nor do I think of conditions of possibility as exclusively external to the phenomena they have bearing upon. Conditions of possibility can be traced internally; that is, they are to be understood as a matter of *constitution*.[23] Specific technoscientific configurations and their conceptual environments enter into and become part of the aesthetic strategies, problems, and matters they help constitute.[24] Stated differently, technology and modernist aesthetics should be understood as *internal* to one another.

I would like to acknowledge one notable omission in this book: the question of gender. I discuss it in places, but it is not a central issue. The relation of gender and technology in literary modernism is a crucial yet strangely undertheorized topic, one that will no doubt change our understanding of the modernist project.[25] The topic would require a separate study, at the very least; it will be a powerful contribution to the revision of modernism now under way.

Proust as Theorist of Technological Change

For a theory of the dialectics of late-nineteenth-century technology and perceptual experience, one could use as a starting point Marx's proposition that the human senses have a history. The cultivation of the five senses, Marx contends, is the product of all previous history, a history whose axis is the relation of humans to nature, including the means with which human beings objectify their labor.[26] One could also draw on the theory of perceptual abstraction implicit in Walter Benjamin's writings on photography, mechanical reproducibility, and Baudelaire's poetry. Benjamin's theory could then be supplemented with Guy Debord's notion of the society of spectacle or, for a more apocalyptic perspective, Paul Virilio's thoughts on the interfaces between technologies of speed and the organization of the human sensorium. One might also consult Marshall McLuhan's theory of means of communication, which, although determinist, usefully suggests that all media can be seen as extensions of the human senses; and Friedrich Kittler's materialist inquiry into the cultural significance of the advent of inscription technologies such as phonography and cinematography.[27]

Yet one would do equally well to begin with Proust. An unsurpassed chronicle of the advent of modern technology, *Remembrance of Things Past* orchestrates a whole world of innovations that parade through the novel from beginning to end, from the telephone to the automobile.[28] What is more, Proust's novel offers a psychology of technological change that may be grasped as a theory in its own right.[29] I am thinking, in particular, of two episodes in *The Guermantes Way* (1920–21), the one revolving around a telephone conversation, the other reflecting on photography. Read together, they offer a sophisticated meditation on the historicity of habits of hearing and seeing. They also serve as a convenient point of departure for distinguishing how such technological change makes available new sensory domains that open themselves to artistic exploration.

The telephone episode relates the narrator's first telephone conversation with his adored grandmother.[30] Transported across vast distances, her disembodied voice hits his ear as though for the first time, "a tiny sound, an abstract sound." The event immediately acquires symbolic proportions:

> It is she, it is her voice that is speaking, that is there. But how far away it is! […] Many were the times, as I listened thus without seeing her who spoke to me from so far away, when it seemed to me that the voice was crying to me from the depths out of which one does not rise again, and I felt the anxiety that was one day to wring my heart when a voice would thus return (alone and attached no longer to a body which I was never to see again), to murmur in my ear words I longed to kiss as they issued from lips for ever turned to dust. (*REM* 2:135/*RTP* 2:432)

For the narrator, the experience is deeply unsettling. But it also awakens the theorizing mind whose speculative brilliance animates long stretches of *Remembrance of Things Past*. Turning around the dissociation of the eye and the ear, of what can be seen and heard, the uncanny event triggers a psychology of telecommunication that stretches over half-a-dozen pages. The narrator realizes that he used to identify what he now perceives as "voice" by matching it with his grandmother's face and other visual features. It is a dialectic moment, for what henceforth appears as having been an organic system of signification has just been sundered; and at the same time, this horizon of signs stands before him, suddenly and visibly revealed, now that it has been lost: "for always until then, every time that my grandmother had talked to me, I had been accustomed to follow what she said on the open score of her face, in which the eyes figured so largely; but her voice itself I was hearing this afternoon for the first time" (*REM* 2:135/*RTP* 2:433).

The narrator discovers not merely his grandmother's voice. Now that he perceives her "without the mask of her face," he also hears, for the first time, "the sorrows that had cracked [her voice] in the course of a lifetime." Dwelling inside her is a figure whom he has never yet apprehended, a figure inhabited by time. The narrator realizes that his grandmother will die, and die soon, and the psychological impact of this insight is irreversible:

"Granny!" I cried to her, "Granny!" and I longed to kiss her, but I had beside me only the voice, a phantom as impalpable as the one that would perhaps come back to visit me when my grandmother was dead. "Speak to me!" But then, suddenly, I ceased to hear the voice, and was left even more alone [. . .]. It seemed to me as though it was already a beloved ghost that I had allowed to lose herself in the ghostly world, and, standing alone before the instrument, I went on vainly repeating: "Granny! Granny!" as Orpheus, left alone, repeats the name of his dead wife. (*REM* 2:137/*RTP* 2:434)

Proust inscribes the telephone episode in the Orpheus myth, thereby reworking the Greek tale in unexpected yet characteristic ways.[31] But there is more to the episode. What the narrator intimates is that a new matrix of perceptual possibilities is sliding into place, one that transforms both the perception of voice (forms of audibility) and the perception of visual appearance (forms of visibility). The experience of the disembodied voice thus elicits a new understanding of that bodily entity from which the voice has been detached.

This, indeed, is confirmed by the episode that follows a few pages later; I shall refer to it as the camera-eye episode. Once these two sections are read in tandem, as I believe they should be, and as I believe Proust meant them to be, an interesting pattern begins to emerge. The narrator is on his way to pay his grandmother a visit, compelled to do so by the telephone conversation and its uncanny revelation of a phantom grandmother, shaded by her age and future death: "I had to free myself at the first possible moment, in her arms, from the phantom, hitherto unsuspected and suddenly called into being by her voice, of a grandmother really separated from me, resigned, having [. . .] a definite age." The narrator enters the drawing room, where he finds her busy reading a book. Because she fails to notice his presence, she appears to him like a stranger. He observes her appearance as he would that of any old woman. In fact, what emerges is that ghostly image that he so desperately wanted to banish from his mind: "Alas, it was this phantom that I saw when [. . .] I found her there reading" (*REM* 2:141/*RTP* 2:438).

The grandmother has become pure image. Why does this stand out to his naked eye? Because she has withdrawn her gaze; it is her failure to look at her grandson that makes him discover, for the second time, her double. During the telephone conversation, her eyes and face failed to accompany her voice, thus anticipating that eternal separation called death. Here, too, she is shrouded in invisibility, for sitting in the sofa is not the grandmother but her double. Disembodied and deterritorialized, she literally emerges as a spectral representation of herself.[32]

In order to explain how the uncanny sight of the grandmother was possible, the narrator draws on the language of photography. He creates an analogy between himself and a professional photographer; he also proposes that during those brief moments before his grandmother realized his presence, his gaze was operating like a camera. The photographic metaphor then sparks a miniature essay on why we perceive our loved ones the way we do, and why these perceptions are always and necessarily faulty:

> The process that automatically occurred in my eyes when I caught sight of my grandmother was indeed a photograph. We never see the people who are dear to us save in the animated system, the perpetual motion of our incessant love for them, which, before allowing the images that their faces present to reach us, seizes them in its vortex and flings them back upon the idea that we have always had of them, makes them adhere to it, coincide with it. How, since into the forehead and the cheeks of my grandmother I had been accustomed to read all the most delicate, the most permanent qualities of her mind, how [. . .] could I have failed to overlook what had become dulled and changed in her, seeing that in the most trivial spectacles of our daily life, our eyes, charged with thought, neglect, as would a classical tragedy, every image that does not contribute to the action of the play and retain only those that may help to make its purpose intelligible. (*REM* 2:141–42/*RTP* 2:438–39)

To drive home his point concerning the alienating vision inherent in the camera, Proust adds yet another example. This scenario, too, rehearses the contrast between what we expect to see, although we may not have realized it, and what we actually perceive:

> But if, instead of our eyes, it should happen to be a purely physical object, a photographic plate [*plaque photographique*], that has watched the action, then what we see, in the courtyard of the Institute, for example, instead of the dignified emergence of an Academician who is trying to hail a cab, will be his tottering steps, his precautions to avoid falling on his back, the parabola of his fall, as though he were drunk or the ground covered in ice. So it is when some cruel trick of chance prevents our intelligent and pious tenderness from coming forward in time to hide from our eyes what they ought never to behold, when it is

forestalled by our eyes, and they, arriving first in the field and having it to themselves, set to work mechanically, like films [*pellicules*], and show us, in place of the beloved person who has long ago ceased to exist but whose death our tenderness has always hitherto kept concealed from us, the new person whom a hundred times daily it has clothed with a loving and mendacious likeness. (*REM* 2:142/*RTP* 2:438–39)

The narrator here splits the category of visual perception into two: the human eye and the camera eye. Marked by affection and tenderness, human vision is necessarily refracted by preconceptions; and such a lens prevents the beholder from seeing the traces of time in the face of a loved one. The beholder sees not the person, merely his or her preconceived images of the person, thus continuously endowing the loved one with a "likeness." Memory thus prevents truth from coming forward.

The camera eye, on the other hand, is cold, mechanical, and undistinguishing. It carries no thoughts and no memories, nor is it burdened by a history of assumptions. For this reason, the camera eye is a relentless conveyor of truth. So it is that the narrator catches sight of a hitherto unknown being, who now flashes into the present: "for the first time and for a moment only, since she vanished very quickly, I saw, sitting on the sofa beneath the lamp, red-faced, heavy and vulgar, sick, vacant, letting her slightly crazed eyes wander over a book, a dejected old woman whom I did not know" (*REM* 2:143/*RTP* 2:440).

The deadly power of the photographic gaze has struck the grandmother who, like Eurydice on the verge of light, instantly vanishes from sight and disappears into the shadows. Left behind is a mere phantom image. To be sure, the narrator's uncompromising image of his grandmother is bound to evaporate as soon as she lifts her eyes and recognizes him. Yet for him those seconds have hinted at her impending death. From now on, her uncanny double is superimposed upon her seemingly everpregiven self.

What starts as a reflection on telephony and the discovery of the disembodied voice thus ends as a meditation on photography and how it changes the perception of visual appearances. In other words, the narrator's effort to grasp the experience of speaking to his grandmother on the telephone motivates a psychology of visual perception as well. Read in this way, Proust offers a germinal theory of how the emergence of technologies for transmitting sound such as the telephone paves the way for a new matrix of perception, in which not only sound but vision also turn into abstract phenomena. McLuhan, in *Understanding Media* (1964), has something similar in mind when he maintains that "the effect of radio is visual, the effect of the photo is auditory. Each new impact shifts the ratios among all the senses."[33]

What is more, Proust's narrator suggests that the perceptual habits of the eye and the ear begin to function separately, each independent of the other, each in its own sensory register. An episode in *Time Regained* (1927) testifies to the consequences of such technological change. Set in the mid-1920s, the scene unfolds at a social gathering. The narrator is reintroduced to an old friend, who expresses delight at meeting again after so many years. A caesura follows, because the narrator, perplexed and confused, fails to identify the person in front of him, although he is able to recognize the voice: "I was astonished. The familiar voice seemed to be emitted by a gramophone [*phonographe*] more perfect than any I had ever heard, for, though it was the voice of my friend, it issued from the mouth of a corpulent gentleman with greying hair whom I did not know, and I could only suppose that somehow artificially, by a mechanical device, the voice of my old comrade had been lodged in the frame of this stout elderly man who might have been anybody" (*REM* 3:985/*RTP* 4:522).

The gentleman's voice, rising out of the body as though of its own accord, is here rendered as a noncorporeal, hence foreign, element. It is the defamiliarizing image of the gramophone that so drastically disconnects the voice from its bodily source. The mechanical metaphor strips the old acquaintance of human qualities such as consciousness and agency, thus reducing him to a nonhuman entity, indeed, to a thing. These images serve to underscore the narrator's insistent efforts to match his perception of the voice with his perception of the friend's exterior; at the same time, they prefigure his utter inability to do so: "I should have liked to recognise my friend, but [. . .] like the visitor at an exhibition of electricity who cannot believe that the voice which the gramophone restores unaltered to life is not a voice spontaneously emitted by a human being, I was obliged to give up the attempt" (*REM* 3:985–86/*RTP* 4:523).

This remarkable episode also turns around the discrepancy between the narrator's aural impressions and his visual experience. What sets it apart from the telephone scenario, however, is that the dissociation of the eye and the ear has already happened. It both precedes and inscribes the narrator's account of the event. If the telephone episode ultimately ponders the spacing of production and reception, of sonic origin and transmission, the present scenario both presupposes and enacts that logic of spacing. Indeed, the representation of the narrator's failure to recognize his friend from long ago is organized precisely by that emerging matrix of perception that the narrator, in *The Guermantes Way*, took upon himself to explain. The representation of the old friend's voice thus presumes the essential internalization of the very experiential effects that the previous episodes set out to chart.

The phonographic metaphor confirms the implicit dialectics at work. In the telephone episode, the narrator contemplated the experience of the pure and abstract voice, intimating that it is enabled by a technology for communicating at a spatial distance. To this sound machine we may now add the phonograph, a mechanical device that makes it possible to strip sound not only of its spatial source but also of its temporal origin.[34] From now on, the voice and other acoustic phenomena are, potentially, subject to endless reiteration and exteriorization.

In this way, then, Proust's telephone and camera-eye episodes articulate a theory of how a new division of perceptual labor comes into play, one that bears both on the habits of the ear and on those of the eye. For although each of these two processes of abstraction may be traced back to its own relatively distinct technological lineage, their experiential effects—reification, autonomization, and differentiation—are fundamentally interrelated. Mutually determining one another, the abstraction of the visual is inherent in the abstraction of the aural, and vice versa.

Meanwhile, as Proust's own phonographic imagery demonstrates, the new optical and acoustic worlds propelled by such change open up realms of representation that readily lend themselves to artistic experiments. From photography to telephony, from phonography to cinematography: technological transformation helps articulate new perceptual domains, charging the modernist call to make the phenomenal world new. Proust's novel thus offers a way of understanding the mediated nature of so many characteristic formal innovations that are to be found in modernist works.

Consider a scene in Woolf's *Mrs Dalloway* (1925). Set in postwar London, the novel invokes a bustling city inhabited by loud airplanes and throbbing motorcars. All of a sudden, an explosion happens in the street, and a vehicle draws to the side of the pavement: "Passers-by [...] stopped and stared, had just time to see a face of the very greatest importance against the dove-grey upholstery, before a male hand drew the blind and there was nothing to be seen except a square of dove grey."[35] It is as though the implied field of vision is delimited by the car window, whose frame then motivates the bringing out of the two distinct visual unities, autonomous and thinglike, which occupy the center of this proto-photographic frame, namely, the face and the hand. The excitement built into the scene derives from the primacy of the visual impression itself, serving to indicate the freshness and sensuous immediacy of the seen.

Perception is heightened in anonymous urban spaces, speculates Ulrich, the hero in Robert Musil's *The Man without Qualities* (1930–32). Exhilarated by the colors of streetcars and automobiles, and absorbed by the shapes of shop windows and façades, Ulrich surrenders himself to the visual stimuli of the city, reveling in "this mass of faces, these movements wrenched loose from the body to become armies of arms, legs, or teeth."[36] Decomposed into a series of free-floating fragments, the visual percept of the crowd indicates that Ulrich's emancipated sense of vision projects its autonomy onto the visual environment that it both envelops and differentiates. That the hero's eyes appear to operate autonomously, endowed with an agency all their own, is then peculiarly affirmed when the narrator likens Ulrich's gaze to an insect. Such a metaphor detaches vision from the human body at one stroke and locates it in a buzzing world of instinct, meanwhile transforming urban space into a landscape of flowering fields, indeed, into nature.

"Artists only see what is necessary to their eye," exclaims the self-styled artist-hero in Wyndham Lewis's *Tarr* (1918), a statement that could serve as a commentary on Lewis's aesthetic method. A giant stylistic experiment, *Tarr* furiously seeks to rewrite the real as spectacle. This is how one of the female protagonists is introduced, the image of her exterior refracted through the eyes of a male artist:

> When she laughed, this commotion was transmitted to her body as though sharp, sonorous blows had been struck on her mouth. Her lips were long hard bubbles risen in the blond heavy pool of her face, ready to break, pitifully and gaily. Grown forward with ape-like intensity, they refused no emotion noisy egress if it got so far. Her eyes were large, stubborn and reflective, brown coming out of blondness. Her head was like a deep white egg in a tobacco-coloured nest.[37]

From Lewis and Woolf to Musil, the world is reinvented in the name of the visual. For all their differences, each description is a unity unto itself, a reified and autonomous entity, working to defamiliarize habitual modes of processing the flux of the real. Such visual scenarios are refracted through a newly emergent perceptual matrix enabled by technologies for transmitting and reproducing the real, acoustic and visual technologies alike. But if such modes of representation are overdetermined by an array of distinct yet interrelated technological phenomena, their aesthetic effect is also, and by the same token, overdetermined. Indeed, for these spectacles to be considered as aesthetically gratifying in and of themselves, there has to be a notion of visual autonomy and sensory differentiation already in place, a notion contingent upon a discursive reconceptualization of perception that, among other things, denaturalizes vision and locates it in the physiological immanence of a human body.[38]

We may glimpse the historical contours of this discursive formation by contemplating briefly the emergence of technologies designed to chart, explore, and record sensory phenomena that had never before been possible to perceive as such. Toward the end of the nineteenth century, voices could be recorded and reproduced at will in all their baffling facelessness; the human skeleton could be visually exposed within the living flesh; and the motion of flying birds and running horses could be translated into stop-motion photographs. As Tim Armstrong emphasizes in *Modernism, Technology, and the Body* (1998), this historical period witnessed "a revolution in perceptions of the body."[39]

In this context, the work of the French physiologist Etienne-Jules Marey is paradigmatic, and I will discuss its implications in the chapter on Proust. Briefly, Marey developed inscription apparatuses capable of turning invisible physiological action into visual records. These devices effectively appropriated the epistemic privileges of the human senses, notably those of the eye. Marey's graphic method and Röntgen's radiographic representations of the human interior therefore share a family likeness. Both initiated new forms of knowledge that simultaneously redefined traditional ones; both contributed to the construction of a visual division of labor that separated what was thought of as the realm of the objective and that of the "merely" subjective; both helped articulate a discursive gap between what was henceforth understood as scientific vision vis-à-vis bodily vision.

Consequently, a powerful discrepancy emerged between, on the one hand, visual means of representing domains that had been inaccessible to the eye and, on the other, the naked eye itself, a discrepancy that provoked a new conception of their respective theoretical tasks. Yet, the very notion of the naked eye, in all its physiological contingency, was an invention, too, since the terms through which it was articulated had been altered. Indeed, the historically strong form of the notion of the naked eye became operative after the successful introduction of technoscientific apparatuses such as those designed by Marey and Röntgen.

If, then, notions of seeing are no longer necessarily linked to categories of knowledge, a dialectical leap must take place. As visual perception is increasingly divorced from knowing, unexplored paths open up; and in the artistic and cultural realm, the inherent value of perception exercised for perception's sake—and, by implication, art for art's sake—gains so much ground as to become something like a modernist imperative.

Such tendencies become increasingly palpable during the second half of the nineteenth century, especially in the realm of the visual arts. It is now that visual perception becomes an aesthetically gratifying activity in its own right. In 1897, Conrad spoke of the purpose of his writing in terms that have long since become famous: "My task which I am trying to achieve is, by the power of the written word, to make you hear, to make you feel—it is, before all, to make you *see*."[40]

Interestingly, Conrad's 1900 novel *Lord Jim* problematizes the power of Western eyes, of dazzled eyes, of all kinds of glances, of the nothing-to-see. It also raises the issue of the freshness of visual perception vis-à-vis the dulling of the eye. This is how Conrad describes the interior of the ship *Patna,* its intermittent light motivating the stylistic decomposition of bodies into so many synecdoches: "A few dim flames in globelamps were hung short here and there under the ridge-poles, and in the blurred circles of light thrown down and trembling slightly to the unceasing vibration of the ship appeared a chin upturned, two closed eyelids, a dark hand with silver rings, a meagre limb draped in a torn covering, a head bent back, a naked foot, a throat bared and stretched as if offering itself to the knife."[41]

Not bodies, but *signs* of bodies. This is how the view of the opaque interior presents itself to the narrator's eye, and Conrad everywhere insists on the priority of the visual impression. Like so many other modernist novels and art practices, *Lord Jim* can be seen as a didactics of subjective, bodily seeing. It is a didactics, moreover, that both builds on and produces the dissociation of knowing and seeing. As we shall see, in both Proust and Joyce the imperative to make the reader see is part of an entire aesthetic program, but with radically different implications.

Seeing and Knowing: From Mann to Joyce

Primarily concerned with sight and hearing, this book puts the spotlight on the two senses that are most immediately affected by the techno-scientific configurations that emerge during the nineteenth century, from photography and phonography to technologies of speed. Each chapter in the critical-interpretive part of the book—chapters 2 through 4—is laid out as an examination of the relations of specific technologies and specific domains of the human sensorium. At the same time, the core of each chapter consists of a reading of a literary text: *The Magic Mountain, Remembrance of Things Past,* and *Ulysses,* each analysis addressing a cluster of notions gathering around the image of technology: body, perception, knowledge, truth, sublimity, legibility, and inscription.

But, as I have already intimated, the trajectory from Mann and Proust to Joyce also tells a story. Each chapter is part of a sequence that casts these modernist narratives as indices of an increasing abstraction of sensory experience. This sequence also traces a certain progression, an ever-closer relationship between the habits of the senses and technologies of perception. It may usefully be grasped as a gravitation from externality toward internalization. All three novels dwell on the issue of how modes of seeing relate to modes of knowing in an age when inscription devices and similar technologies claim superiority as means of registering and storing sense data; and this issue, I argue, has immediate bearing on the aesthetic strategies these writers articulate and explore. Yet although these modernist narratives respond to and enact the crisis of the senses, both probing and staging the dissociation of categories of knowing and seeing, the differences between them nevertheless merit careful consideration, not so much because of the differences in subject matter, but rather because of the radically different terms in which it is articulated.

What, then, happens when scenarios of seeing are no longer necessarily related to scenarios of knowing? And to the extent that the novel used to be seen as an instrument of humanist and therefore "universal" knowledge, how does such a transformation affect the status of the novel? As I discuss in Chapter 2, these questions penetrate to the core of *The Magic Mountain,* a narrative that is obsessed with the eye and the limits of what it may see and know. Subjected to an unusual form of aesthetic education, Mann's hero is confronted with a series of visually signifying systems that relativize one another, from the visually based hermeneutics of class and the structured perception of the medical gaze to the mechanical eye of the X-ray apparatus. Each system corresponds to a mode of interpretation, each producing different answers to the questions around which the narrative revolves: What is a human being? What is identity? How can we know what we know? What is the role of art and culture? Yet this is only part of the story, for *The Magic Mountain* does more than dramatize the ways in which the powers of the human eye are marginalized. It also links this epistemic crisis to the theme of *Bildung* and, by implication, to subjectivity. If Mann's novel thus emerges as an inquiry into the *epistemic* mandates of the human eye, then Proust and Joyce work in the opposite direction: *Remembrance of Things Past* and *Ulysses* interrogate the *aesthetic* powers of the eye. Mann's emphasis, one might say, is on the first term of the knowing/seeing dissociation, Proust's and Joyce's on the second.

In *Remembrance of Things Past,* as I have intimated, the advent of modern technology is part of the subject matter, fueling the great theme of lost time. But technological change is also inscribed in the inner form of the narrative, and nowhere more so than in Proust's aesthetics of visual perception. In Chapter 3, I trace how chronophotography and cinematography and their inherent matrices of perception refract vital parts of the aesthetic sensibility for which *Remembrance of Things Past* serves as a vehicle. Proust's novel tells the story of the making of a man of letters, and according to a commonly adopted view, it is only when the past claims the unsuspecting narrator through his deep-seated bodily memories that he may become a writer. Yet not all of Proust's world rises out of a teacup. The emergence of writing is also closely connected to technologies of velocity and the new visual worlds they burst open. One of the earliest writers to elaborate a syntax of speed, Proust enters an optical universe flung open by technologies of velocity, such as the motorcar, and articulated by technologies of vision, which include chronophotography and cinematography. For this and other reasons, I argue, the author of *Remembrance of Things Past* reveals himself as a prototypical modernist, remaining true to John Ruskin's imperative: to keep to what one perceives rather than what one knows.

This is certainly true of the author of *Ulysses,* too, but Joyce emerges as a special case. If both Mann and Proust explicitly engage the dissociation between seeing and knowing, then in Joyce this divorce has already taken place. In fact, the call to *perceive* nowhere makes itself so strongly felt as in *Ulysses.* Joyce's modernist epic turns Conrad's imperative to make the reader feel, hear, and see into an axiomatic and autonomous aesthetic principle. In Proust, after all, perceptual activity ultimately yields new and different forms of knowledge, whereas in Joyce perception has become an end in itself, disconnected from the accumulation of knowledge. In Chapter 4, my analysis focuses on the ways in which Joyce represents visual and aural sensations. This, it turns out, is to trace a set of stylistic procedures that, despite the diversity that is the novel's hallmark, persist throughout *Ulysses.* Joyce's style pursues the everyday in all its lived minutiae, entering so deeply into the particular that the thing-in-itself, in all its wished-for freshness and irreducibility, appears to impose itself on the reader: a teaspoon, a sausage, a fingertip, a tapping cane, a clacking tongue, a smelly onion. Joyce's zest for the particular is nevertheless a historically specific will-to-form. Indeed, what I designate as Joyce's aesthetics of immediacy can be seen as a stylistic attempt at solving a historical problem, for in order to render the freshness and concreteness of the perceived, Joyce appropriates matrices of perception inherent in cinema, phonography, and telephony, turning them into techniques for rendering the immediacy of lived experience in an age where the very category of experience is subject to debate.

Where the reading of *The Magic Mountain* pays attention to narrative content and generic form, and where the analysis of *Remembrance of Things Past* focuses on thematic structures and questions of style, the chapter on *Ulysses* dwells on stylistic inflections at the level of the sentence. This methodological trajectory is part of the larger story: the ever-closer relationship between the habits of the sensorium and technologies of perception, including the tendency toward internalization. Because the processes of technological change thematized by the novels of Mann and Proust have been fully incorporated in Joyce's novel, they are to be detected in the stylistic, syntactic, and rhetorical dimensions of *Ulysses,* and in them alone. In effect, Joyce's style registers the subterranean effects of those technological inventions that Mann and Proust subject to discussion. Yet the Joycean sentence does not only reveal the increasing dissociation of modes of perception and modes of knowing. It even amplifies this same process, for with *Ulysses,* to see for the sake of seeing and to hear for the sake of hearing become aesthetic ends in themselves. So it is that *Ulysses* represents a certain completion of the inherent tendencies that may be traced in *The Magic Mountain* and *Remembrance of Things Past:* the marginalization of the epistemic mandates of the human senses in an age where technological devices increasingly claim sovereignty over and against the sensorium. Joyce's novel thus represents the full-blown internalization of technological modes of reproducing the real.

If, as I shall argue in the pages that follow, it can be shown that the operations of a variety of technoscientific configurations are inscribed in the specific idioms, procedures, and innovations that have commonly been seen as representative of high-modernist aesthetics, then the terms in which traditional historiographies of modernism are couched will have to be reconsidered, particularly the notion of aesthetic autonomy, but also the influential presupposition that there is a divide between a merely technological culture and a more properly aesthetic one. The ethos of high-modernist aesthetics has for so long been considered alien to everything technological, utilitarian, and rationalized that this opposition has become a conceptual commonplace and historiographic design in theories of modernism. It is time to ask how and why this founding myth of modernism has emerged and persisted. As Ezra Pound, who probed the intersections of *poeisis* and *techne,* once put it: "You can no more take machines out of the modern mind, than you can take the shield of Achilles out of the *Iliad.*"[42]

1 The Antitechnological Bias and Other Modernist Myths

Literature and the Question of Technology

Shortly before the outbreak of the Great War, the French intellectual Charles Péguy remarked that the world had changed more in the last thirty years than it had since the death of Christ. In a similar spirit, Louis Aragon remarked upon the pulsations of the new. "Each day alters the modern feeling of experience," he wrote in *Le paysan de Paris* (1926), translated as *Nightwalker.* "A mythology takes shape and comes undone."[1] These observations epitomize the self-consciousness of those generations of Europeans whose historical experience spanned the last decades of the nineteenth century and the beginnings of the twentieth. They witnessed the emergence of that phase of modernity sometimes called big-scale or imperialist capitalism, a historical period in which the new was fated to be out of date at a pace previously unseen.

In the postmodern age, the kind of temporal and historical experience upon which Péguy's and Aragon's observations build is no longer applicable. New perspectives emerge and modernist culture lays itself open to historical reflection. Not everyone might agree that we live in a postmodern era, to be sure, but probably no one will dispute the fact that modernist literature, painting, sculpture, film, music, and architecture have long since, and for a variety of reasons, entered the halls of tradition. Clearly, it is no longer self-evident that "Modernism is our art," as Malcolm Bradbury and James McFarlane claimed in 1976, when they published their magisterial anthology *Modernism, 1890–1930.*[2] If the modern world increasingly invites an anthropological perspective, this is because something in our conception of it has changed.[3] It has become possible, in other words, to approach modernist texts with a historical perspective, where high modernism—its practitioners and advocates, its texts and artifacts, its sensibilities and programs—is understood not as it wanted to understand itself, that is, as autonomous aesthetic performances at a safe remove from the encroachments of technology, massification, and commodification, but rather as historically mediated, semiautonomous cultural practices responding to, expressing, or managing conceptual, ideological, social, and cultural crises. The following pages assume that from the viewpoint of the postmodern—some would call it a privileged viewpoint, others a perverted one—modernism is bound to look different. To paraphrase Harold Bloom's formula, a postmodern vantage point yields a productive misreading of modernism or, which amounts to the same thing, a strong reading.[4]

It is therefore all the more interesting that Slavoj Žižek has located the break between modernism and postmodernism in the "inherent relationship between the text and its commentary." Žižek's argument can be fleshed out as follows. A modernist text is difficult, complex, and more or less incomprehensible. Its ambitions are monstrous; in between the covers of a single work, the writer wanted to encompass all of the universe, from the toils of the average person to the heights of philosophy, as in Mann. More often than not, the intention was to estrange the habitual perceptions of daily life, thus proposing a more rigorously aesthetic attitude toward a cup of tea, as in Woolf and Proust. If the modernist text appeared incomprehensible—consider Joyce's *Ulysses* and *Finnegans Wake*—the business of the reader or the critic was to "complete" the work of art by revealing its inner continuity with the practices of daily life, that is, to familiarize the seemingly unfamiliar. "If, then, the pleasure of the modernist interpretation consists in the effect of recognition which 'gentrifies' the disquieting uncanniness of its objects," Žižek writes, "the aim of the postmodernist treatment is to estrange its very initial homeliness." Postmodern interpretation thus tends to focus on cultural products with a deliberate mass appeal, detecting in them an "exemplification of the most esoteric theoretical finesses of Lacan, Derrida or Foucault."[5]

To qualify Žižek's suggestion: not only the status and nature of the interpretive act are at stake, but also the question of value. Until fairly recently, the task of literary scholars was to evaluate works of art, to assess whether they conformed to certain aesthetic standards, and ultimately, to determine whether they deserved a place in the literary canon. With the arrival of a more properly modernist culture and the related academic institutionalization of modernism in the 1950s, the problem of evaluation ceased to be a central concern for scholars.[6] The question of value had become implicit; value was, as it were, embedded in the interpretive act itself, an activity that ensured the handing down of the work of art to new generations of readers.

With the emergence of postmodern culture, the question of value is again brought to the surface, but with a difference. If a scholar penetrates a rock video in excruciating detail, it is not necessarily because it is deemed valuable in the older sense, but because it is seen as the occasion for, say, a discursive analysis. The question of value thus returns, but as an institutional symptom that points to the proliferation of incompatible "value spheres" within the academy, to use Max Weber's term, each with its own standards and norms, each with its own interpretive practices and rationales. The very fact that Žižek is capable of juxtaposing a modernist hermeneutic model with a postmodern one testifies to the relativization of modernist reading practices and, by implication, modernist aesthetics.

As the modernist value system is losing legitimacy, and as new interpretive practices and rationales emerge in its wake, one may expect new reading methods as well. This in turn is likely to entail that the landscape of literary modernism, as we know it, will be redrawn. What, then, happens to the supposedly constitutive features of modernism once they are looked upon from the perspective of postmodernist culture? A growing number of scholars are probing the fact that the modernist era is also the period of women's movements and labor movements, of class conflict and revolutions, of scientific discoveries and technological change.[7] The modernist period also coincides with the final stages of the age of empire, when the capitalist system expanded into a fully developed imperialist system. Edward Said, in *Culture and Imperialism* (1993), suggests accordingly that "many of the most prominent characteristics of modernist culture, which we have tended to derive from purely internal dynamics in Western society and culture, include a response to the external pressures on culture from the *imperium*." Commenting on Mann's novella *Death in Venice* (1912), Said proposes that Venice, located as it is on the Asian border of Europe, is no mere background for Aschenbach's spiritual crisis, but part of the logic of the plot. Taken as a whole, the novella intimates that "Europe, its art, mind, monuments, is no longer invulnerable, no longer able to ignore its ties to its overseas domains."[8] Read as an allegorized Eurocentric fantasy, *Death in Venice* no doubt resonates with all the troubling geopolitical insight that Paul Valéry was to articulate in a 1919 essay with a now famous opening: "We others, civilizations, we now know that we are mortal."[9]

Against this background, new ways of reading European modernism suggest themselves. Yet the immense task of exploring systematically the formal and structural traces of those "external pressures" in modernism is one that still awaits its theoreticians, scholars, and critics. The first question is perhaps this. To what extent, if at all, are these pressures to be thought of as external?

The Myth of the Split

Modernism, as literary historians agree, is a notoriously imprecise term. It merely signifies the empty flow of time. One does not have to be a fervent nominalist to realize that an unbridgeable gap exists between the term "modernism" and the vast array of phenomena—cultural, social, even economic—to which it is taken to refer. Yet even a cursory glance at the vast secondary literature on that heterogeneous phenomenon known as modernism shows that certain conceptual patterns, commonplaces, and topoi tend overwhelmingly to recur. One example is the opposition between the old and the new. Numerous accounts turn on the notion of the now and the new, since *modernus,* as an adjective and a noun, originates in the late Latin use of the adverb *modo,* "recently, just now."[10] Indeed, most expository studies of literary modernism open with a reflection upon the word "modernism" itself, sometimes followed by a survey of the history of the seemingly recurrent quarrel between *les anciens* and *les modernes.*[11] Modernism can then be seen as the expression of an underlying proto-historical logic that has erupted fairly regularly (since, say, the sixth century) in the many collisions between the old and the new. In accounts of these changes, the questions at issue tend to be both raised and answered in terms of style, aesthetics, or similar "internalist" notions.

Another conceptual pattern that frames a majority of surveys of literary modernism is the opposition between the social and the aesthetic. This opposition frequently assumes the form of a dualism between technologized mass culture and nontechnologized high culture.[12] I shall have more to say about this later. What is significant here is that almost all expository studies of modernism and quite a few monographs turn on a series of familiar and recurring oppositions that are generated by, related to, or congruous with those of art and society. A quick catalogue of these oppositions, where the valorization of the first term invariably presupposes the marginalization of the second, might read as follows: art/commerce, art/instrumentality, art/industry, aesthetic discourse/communicative discourse, art/market, aesthetic value/economic value, beauty/utility, autonomy/reducibility, innovation/standardization, originality/reproducibility, high culture/mass culture, body/machine, organic/mechanical, individual/collective, minority/majority, elite/masses, and so on. My point here is not so much that such dualisms exist in modernist studies—they are, after all, common enough. My point is rather that their specific topographies, interrelations, and distributions of value are so stabilized as to amount to something like a discursive system. This system allows the critic to construct a vision of history which then serves as a model for situating modernist literary practices within a larger historical context.

I do not intend to chronicle here the existing mass of scholarly works on modernism. Instead I will select a few representative studies, mostly expository ones, that aim explicitly to narrow down the nature of modernism, either as a recognizable stylistic project or as a set of aesthetic movements. My typology is based on how the critical paradigm in question theorizes, explicitly or implicitly, the relation of literary modernism to history, including that of modernism to technology. Three major approaches can be distinguished: the essentialist, the humanistic, and the historical-materialist approach. These critical paradigms are certainly not mutually exclusive; in fact, they often overlap. The following survey is therefore to be seen as a discussion of dominant tendencies within modernist studies.[13]

In the essentialist approach, the notion of autonomy is paramount. Critics working within this paradigm operate with the assumption that essentially modernist traits and features do exist, whether stylistic, semantic, or rhetorical, and that they can be identified and typologized. Douwe W. Fokkema's study *Literary History, Modernism, and Postmodernism* (1984) testifies to the persistence of the essentialist approach. Setting out to define the typically modernist "semantic universe," Fokkema aims to distinguish it from its postmodern counterpart.[14] The crucial question of why modernism would be replaced by postmodernism is left unanswered, however, as Fokkema's spatialized description of the nature of the modernist text builds upon a formalist notion of meaning and, like most formalisms, fails to account for the diachronic aspects. As a result, the category of history, even if only in terms of "influence" or "context," is precluded.[15]

A second approach—here designated as humanistic—tends to emphasize the continuities of, for example, "human nature," "universal values," "art," and "beauty," cultivating that which seems to have existed for all time. Although the notion of autonomy is important here, too, critics maintain that the nature of aesthetic modernism must be understood against a historical background. The word "modern" appears to suggest a story, a reflection upon what went before it: "tradition," "the premodern," and the like. Here, the relations of literature and history, literature and society, or literature and science are mostly conceived of in terms of "contexts" and "influences." What characterizes the humanist approach is a particular conceptual pattern, namely, the opposition between the planned, rationalized, and predictable on the one hand and the artistic, imaginative, and nonreducible on the other. In fact, it is possible to discern the workings of something like a myth—more specifically, the myth of a split between the social and the aesthetic. As we shall see, it serves to found not only a literary typology but also a rudimentary historical narrative.

Hugo Friedrich's widely influential study *The Structure of Modern Poetry* (1956) delineates the trajectory of modern poetry from Baudelaire onward. Combining the essentialist approach with a traditionally humanist perspective, Friedrich sets out to identify specifically modernist stylistic techniques, rhetorical figures, and syntactic structures. At the same time, he attempts to situate these features in their historical context, that of modernity. In Friedrich's view, modernity is characterized by dehumanization, massification, urbanization, positivist science, and utilitarian language. His question then becomes: How is poetry possible in a commercialized and technologized civilization? To be sure, Friedrich usefully emphasizes that numerous modern poets, notably Charles Baudelaire, Arthur Rimbaud, and Guillaume Apollinaire, were ambivalent toward modernity. But his inquiry is nevertheless organized around the myth of the split, and poetic imagination is placed in contradistinction to technologized civilization:

> Poetry is haunted by the suffering from lack of freedom in an age [. . .] in which the "second industrial revolution" has reduced man to a minimum. He has been dethroned by his own machines, the products of his power. [. . .] The flight into the unreal, the abnormal imagination, the deliberate mysteriousness, the hermetic isolation of language: all these can be understood as attempts of the modern soul, trapped in a technologized, imperialistic, commercialized era, to preserve its own freedom as well as the miraculous in the world.

Once the pattern of this historical development has been established, Friedrich mobilizes a qualifier that sets modern poetry apart from earlier lyrical traditions: "In the modern age, the world issuing from creative imagination and sovereign language is an enemy of the real world."[16] As in numerous expository studies of literary modernism, Friedrich's organizing distinction—between the planned, rationalized, and predictable on the one hand and the artistic, imaginative, and nonreducible on the other—serves both as a way of characterizing aesthetic modernism and as a literary historical mode of apprehension.

A similar oppositional structure frames the art critic Clement Greenberg's programmatic essay "Avant-Garde and Kitsch" (1939). Greenberg here seeks to articulate the nature of kitsch, or mass culture, and to defend its alleged other: the avant-garde or, which amounts to the same thing, high culture.[17] For the first time in the history of Western art, he argues, it has become difficult to distinguish the specific values that can be found in art alone. In an effort to locate the cause, Greenberg contends that kitsch has blurred the boundary between art and nonart: "Kitsch, by virtue of a rationalized technique that draws on science and and industry, has erased this distinction in practice." The underlying opposition is familiar enough, as is its distribution of value: Greenberg places (nontechnologized) art in stark contrast to (technologized) kitsch, meanwhile ascribing to the former an intrinsically high value that the latter can never possess. At the same time, however, Greenberg believes standards of taste and value to be a function of class and education. He is therefore forced to argue that the realm of art is an exception to the relativity of standards of taste, lest his basic opposition between avant-garde and kitsch cannot be sustained. Yet he does not mobilize arguments that would support his thesis about the *inherent* qualities of avant-garde art or high culture. He evokes instead the existence of a historical consensus that is nothing but the external sign of the absolute standards of taste with respect to art. "Taste has varied, but not beyond certain limits," he claims accordingly, for "there does seem to have been more or less of a general agreement among the cultivated of mankind over the ages as to what is good art and what bad."[18] Why, then, is kitsch bad taste? First, because kitsch, or mass culture, appeals to the many and, second, because its mode of production is technological. And since it is technologically produced, it can have no intrinsic qualities. Kitsch is bad because it is bad, and this tautological logic is a byproduct of the opposition between the "technologized" and the "nontechnologized."

Such a conceptual pattern recurs in numerous studies of modernism. Consider, for example, Matei Calinescu's comprehensive and learned survey, *Five Faces of Modernity: Modernism, Avant-Garde, Decadence, Kitsch, Postmodernism* (1987). Calinescu's foundational premise is this. There are "two distinct and bitterly conflicting modernities," the one bourgeois, the other cultural. During the first half of the nineteenth century, "an irreversible split occurred between modernity as a stage in the history of Western civilization—a product of scientific and technological progress, of the industrial revolution, of the sweeping economic and social changes brought about by capitalism—and modernity as an aesthetic concept."[19] This split, however, is never argued in rigorously historical terms. Calinescu nevertheless proceeds to use this divide as the basis for a historiographic model, rudimentary though it may be. In other words, the assumption that a gap exists between bourgeois and cultural modernity thus turns into a justification for studying the latter—avant-garde, kitsch, and decadence—as a mere reaction to bourgeois developments.

In Calinescu's scheme, yet another split exists, and it cuts through cultural modernity, separating mass culture, or kitsch, from high culture. As in Greenberg, the opposition between kitsch and high culture turns into an implicit axiology: mass culture is inherently bad, mediocre, and degraded; high culture inherently good, superior, and boundless. Mass culture is so heavily imbricated in the values and beliefs of middle-class life that it cannot claim the word "culture" for itself, while high culture is believed to transcend these conditions.[20]

Calinescu's explanatory model builds upon the opposition between the aesthetic on the one hand and the social, technological, industrial, and rationalized on the other. This means, among other things, that the attempt to explain the emergence of aesthetic modernity easily resorts to nonhistorical explanatory models. History is thus readily exchanged for psychology: "What defines cultural modernity is its outright rejection of bourgeois modernity, its consuming negative passion."[21] Similarly, cultural modernity can be theorized in terms of temperament; it was "disgusted with the middle-class scale of values and expressed its disgust through the most diverse means, ranging from rebellion, anarchy, and apocalypticism to aristocratic self-exile." Clearly, then, Calinescu's point of departure—the split between cultural and bourgeois modernity—inevitably marks the historiographic structure of the larger story told: how cultural modernity seeks to transcend, negate, or explode certain historical developments and turns into the other of bourgeois modernity.

A similar circular logic is at work in Ricardo Quinones's *Mapping Literary Modernism* (1985). An inventory of aesthetic sensibilities, Quinones's expository study seeks to reconstruct the self-understanding of the modernists. He maintains, first, that modernism is marked by, in his terms, a split between public time and subjective time, and second, that in affirming subjective time, this feature must be understood in relation to renaissance culture, where time was "predictive as well as innovative." According to Quinones's historical scenario, renaissance developments and the related predictive aspect of time eventually result in the "industrial mind of the nineteenth century" and the "standardization of experience." This industrial culture then creates its own dissidence: modernist art. In accordance with the psychological explanatory model, Quinones maintains that the modernists, unlike the futurists, "did not feel a need to identify with the forward movements of industrial society. They felt the end-products to result in stagnation, showing no real change based upon heterogeneity and variety, but rather repetitive sameness."[22] Thus literary modernism, protesting against the "eclipse of the innovative," goes into exile.

Whether one believes that cultural modernity must be understood as a renunciation of bourgeois modernity, as Calinescu does, or that the modernists wanted to withdraw into self-imposed exile, as Quinones does, the conceptual pattern is the same. Both propositions amount to the same view: the condition of modernist culture is opposition. As a consequence, aesthetic modernism tends to be understood as external to modernity. Clearly, it is necessary to search for an alternative conceptual model, one that allows for a rearticulation of the notion of autonomy so that modernism can be seen as a socially significant set of cultural practices in the realm of literature. This is a problem that Astradur Eysteinsson addresses in his penetrating inquiry into paradigms of literary modernism, *The Concept of Modernism* (1990). Critiquing a major conceptual tendency in modernist studies, Eysteinsson asks how "the concept of autonomy, so crucial to many theories of modernism, can possibly coexist with the equally prominent view of modernism as a historically explosive paradigm."[23] Accordingly, he challenges critical models that rely on an aesthetics of autonomy, particularly the notion that a specifically literary language exists. He urges that literariness is always dependent on the context in which the speech act or linguistic artifact is produced or received. Any inquiry into the nature of a specifically modernist literary language thus has to account for its sociolinguistic context. In the end, he concludes, those cultural practices that go under the name of modernism defy exclusively formal or stylistic definitions.

Eysteinsson also scrutinizes periodizing theories of modernism and accounts of modernism as a historically specific phenomenon, from Wellek and Warren to de Man, from Adorno to Bürger. These approaches, no matter how philosophically sophisticated, are also found wanting, primarily because they take for granted the intrinsic value of high culture. In an effort to reconcile the notion of modernism as a culturally subversive force with the notion of modernism as an aesthetic-formalist project, Eysteinsson proposes that literary modernism be rephrased as an "aesthetics of interruption," that is, as an ensemble of negative linguistic practices that attempt to open up a "critical space" in the realm of "instrumental rationality" and "processes of habitualized communication."

Despite its emphasis on meaning as a context-bound phenomenon, Eysteinsson's study pays little attention to the question of what the context of modernism might be. Furthermore, the alternative model posed by Eysteinsson, although incisive and theoretically grounded, stops short of transcending the conceptual horizons of those paradigms of modernism he set out to critique in the first place. As long as the problem at issue is raised, framed, and discussed in these terms—that is, as autonomy versus context—the solution will also, and inevitably, be confined to the very same terms, including the oppositional structure through which they are framed. As a result, Eysteinsson's model can only produce yet another version of the split between the aesthetic and the social, between modernism and history.

Instead of asking how the notion of autonomy can be reconciled with the argument that meaning is context-bound and situation-specific, it seems a more pressing task to ask why and how this opposition has become a problem. Only then does it become possible to challenge the conceptual topoi that organize the influential paradigms of literary modernism that have been handed down to us.

The Manufacture of Art: Autonomy and Instrumentality

The advent of new communication technologies in the nineteenth century had an obvious impact upon the production of art, as had the emergence of large readerships and groups of consumers.[24] Modernist aesthetics and the notion of the autonomous work of art are indirectly related to the same process. During the modernist period, the concept of autonomy is reinforced and reconfigured in crucial ways, first with the *l'art-pour-l'art* movement, then with symbolism and later with various avant-garde movements, all the way up to Frankfurt School cultural criticism and Anglo-American formalist humanism, including New Criticism. As I have already suggested, the notion of aesthetic autonomy coincides in certain ways with the idea that high-cultural art is more deserving of scholarly attention because, among other things, it is less informed by technologized production, and by the same token, that mass culture, as a low-cultural phenomenon, is less deserving of scholarly attention since it is far removed from the activity of artistic imagination. Generally speaking, the modernist period is marked by an ideology of a partition—between technology and its effects on the one hand, and the ostensibly free activity of the artistic mind on the other. One would have to go back to preromantic times in order to find a notion of art and of aesthetic activity that does not operate in contradistinction to *techne*.[25] Underlying the topos of the split is the idea that technology equals streamlined industrial production, utilitarian logic, or instrumental reason. In other words, the nature and operations of technology are subsumed under the category "means," as in "means to an end."

This becomes particularly evident in Adorno and Horkheimer's reflections on the culture industry in *Dialectic of Enlightenment* (1944). Here, too, the explicit opposition between mass culture and modernist art coincides with a distinction between inauthentic and authentic art. But the more important point is that the related distinction between the technologized and the nontechnologized organizes the critique of the culture industry as a whole. For Adorno and Horkheimer, order, unity, and rational planning are the fundamental principles of mass culture. The producer is the divine creator of the universe, the master builder whose all-encompassing consciousness leaves its mark on the most minute part of the mass-cultural artifact: "The whole and the parts are alike; there is no antithesis and no connection. Their prearranged harmony is a mockery of what had to be striven after in the great bourgeois works of art."[26] Reducing the parts to a preconceived and therefore false whole, the producers of mass culture cast the heterogeneity of potentially "artistic" material into the mold of the same. Thus, mass culture both testifies to and contributes to the power of instrumental reason. High-modernist art, on the other hand, transcends the same historical laws, capable of securing for itself a realm from which it is possible, in negative fashion, to launch a critique of these laws. In Adorno and Horkheimer, the oppositional structure of the general argument nevertheless forms the basis for the implicit question of value. The presumably nontechnologized is placed in stark contrast to the technologized, and mass culture is, once again, inherently bad and inauthentic. The problem, it seems, lies in the conception of the notion of autonomy. A brief sketch of its emergence helps explain why this is so.

The common, essentially romantic understanding of the meaning of words such as "art" and "culture," at least in the English context, emerged, as Raymond Williams has claimed, both as a result of and as a critique of the dramatically increased process of industrialization during the eighteenth century. The production of art progressively had to resort to the laws of the market rather than those of the church and the aristocracy. As the patronage system was abandoned, poetry and prose turned into commodities in a literary market proper. During the course of the eighteenth century, the meaning of the word "art," which commonly had meant "skill," developed a specialized significance, first in the realm of painting, then in that of the imaginative arts more generally. By the same token, a new distinction emerged between "artist" on the one hand and "artisan" and "craftsman" on the other. In the world of art and imagination, as opposed to that of craft and fancy, the emphasis on skill turned into an emphasis on "sensibility," "creativity," and "originality."[27]

If, as Williams suggests, the construction of the romantic artist and the thought of the ideal implied reader as one who admires "what delights the cultivated" were mediated by the emergence of a literary market proper, one might also argue that the conception of the romantic artist built upon an already available discourse, where the oppositions between the organic and the mechanical, the human and inhuman, the original and the imitative had established themselves as readily comprehensible conceptual designs. If technology had commonly been seen as an extension of the human body or as a matter of skills, it now began to acquire a new significance; indeed, technology was henceforth understood in stark contrast to everything human and corporeal. Thus Edward Young, writing in 1759, can be said to reproduce in the realm of aesthetic theory a socially recognizable opposition between art and *techne*: "An original may be said to be of a vegetable nature; it rises spontaneously from the vital root of genius; it grows, it is not made: imitations are often a sort of manufacture wrought up by those mechanics, art and labour, out of preexistent materials not their own."[28]

Young's oppositions arrange themselves neatly into two basic conceptual compartments: nature and culture, or the natural and the artificial. Thus Young reinscribes the romanticist invention of the idea of Nature as virgin land, an idea that can be found already in Milton's *Paradise Lost* (1667). In Milton nature is allegorized as a vast female body, contrasting sharply with the imagery of mining, which, furthermore, is projected onto the image of hell: "There stood a hill not far, whose grisly top/Belch'd fire and rolling smoke; the rest entire/Shone with a glossy scurf; undoubted sign/That in his womb was hid metallic ore,/The work of Sulphur."[29] Here as elsewhere, Nature emerges as an untouched island. As a negative afterimage of an increasingly industrialized society, Nature must henceforth come down on the side of the human, as must indeed Art.[30]

Aesthetic discussions in late-eighteenth-century Germany display similar tendencies. The publication of Kant's *Critique of Judgment* (1790) was the crowning achievement, of course, but as Martha Woodmansee argues in *The Author, Art, and the Market* (1994), the Kantian notion of the autonomous and disinterested work of art was preceded by decades of aesthetic debate in journals, newspapers, private letters, and legal documents, involving Gotthold Ephraim Lessing, Friedrich Schiller, Karl Philipp Moritz, Moses Mendelssohn, and others. In these debates, an older theory of art—Woodmansee designates it as instrumentalist—was opposed to the emergent idea of the *innere Zweckmäßigkeit,* or internal perfection, of the work of art conceived as a harmonious, self-sufficient whole. Situating the emergence of Kant's aesthetics in a network of discourses that referred to several mental spaces all at once, philosophical, cultural, legal, and economic, Woodmansee demonstrates that the debates on instrumentalism versus autonomy were not only part of the ongoing discussion concerning the status of the artist and his work—a discussion which, by implication, also concerned the regulation of the relationship between writer and reader. The debates were also integral to the legal discussion about intellectual property, which eventually led to copyright laws.[31] The move from a pragmatics of art to a conception of internal perfection was paralleled by a move from the understanding of the book as a physical medium for presenting preordained truths to the notion of the book as an emanation of the intellectual and creative capacities of a unique individual. Only then, Woodmansee maintains, was it possible for the author to claim ownership of the products of his labor.

One might therefore say that the invention of the space of the genius, a space constructed in contradistinction to the market, coincided with the invention of the notion of aesthetic autonomy and the exclusive focus on the *innere Zweckmäßigkeit* of the work of art—a notion that Kant was to develop into *innere Zweckmäßigkeit ohne Zweck.* Indeed, as Adorno and Horkheimer underscore in *Dialectics of Enlightenment,* the "principle of idealist aesthetics—purposefulness without a purpose—reverses the scheme of things to which bourgeois art conforms socially: purposeless for the purposes declared by the market."[32] Stated simply, literary aesthetics from Kant onward thus bears the traces of the social and economic situation in which the notion of autonomy emerged. It is mediated by, among other things, the conditions of commercial publishing and, by im-

plication, the industrialization process. As Roger Chartier has emphasized, although professional writers and authors were dependent upon direct monetary compensation, their self-understanding nevertheless built
up the idea of the radical independence of their products. Indeed, the
shift "was accompanied by an apparently contradictory change in the ideology of writing, henceforth defined by the urgency and absolute freedom
of creative power."[33]

In this familiar sketch of the historical conditions of the notion of aesthetic autonomy, technology, encapsulated by the word "industrialization," has figured in the conceptual periphery. Yet this account should
make it obvious why a majority of theories of modernism have come to
see modernism in terms of a split between, on the one hand, authenticity,
autonomy, and art; and on the other degradation, technology, and massi-
fied culture. Such theories understand modernism largely as it wanted to
understand itself. But, as has already been suggested, while it is no doubt
true that modernist cultural practices often sought to transcend, critique,
or otherwise undermine what was thought of as technoscientific ideology,
instrumental reason, commodification, and so on, modernist art and culture were nevertheless profoundly affected by those same phenomena.

What deserves emphasis here is that the conceptual features I have
been tracing—the myth of the split, including the dualism between modernism and modernity, as well as the standard axiologies—are vital components of the discursive system through which literary modernism has
been understood and theorized. What also deserves emphasis is that in
this binaristic discourse, the concept of technology invariably appears in
the same negatively charged cluster as science, instrumental reason, and
utilitarian thought. Another example may be drawn from Renato Poggioli's *The Theory of the Avant-Garde* (1962). What often reigns supreme in
avant-garde art and modern culture at large, Poggiolo asserts, is "technicism," that is, "the reduction of even the nontechnical to the category of
technique." Technicism, he argues, "means that the technical genius invades spiritual realms where technique has no raison d'être. [. . .] It is
not against the technical or the machine that the spirit justly revolts; it is
against this reduction of nonmaterial values to the brute categories of the
mechanical and technical."[34] Given the prevalence of such antitechnological bias, it is hardly surprising that the question of the relatedness of art
and technology has been omitted from a majority of modernist studies.
But, as I will insist throughout, we dissociate the question of aesthetics
and that of technology at the cost of not recognizing the complexity, richness, and hybridity of modernist cultural practices.

The conceptual patterns I have been discussing also tend to recur in theories of modernism that are explicitly historical-materialist or dialectical. As we saw above, Adorno and Horkheimer's discussion of the culture industry is a case in point. The correspondence between Adorno and Benjamin in the mid-1930s provides another example. Debating the status of the work of art in the age of mechanical reproduction, Adorno weighed the possibilities of the (autonomous) great work of art against those of (reified and therefore degraded) mass culture. He objected to Benjamin's enthusiasm for the emancipatory potential of cinema and his coldness towards *l'art pour l'art*, but nevertheless admitted that "both bear the stigmata of capitalism, both contain elements of change [. . .]. Both are torn halves of an integral freedom, to which they however they do not add up."[35] Adorno's negative dialectics, however, tipped the scale in favor of high modernism, particularly the European tradition of experimental writing stretching from Stéphane Mallarmé to Kafka and Samuel Beckett. Such modernist writings, Adorno maintained, were inherently capable of delivering a subversive critique of the very social processes that had enforced their autonomization.

By delineating the adventures of the topos of the split between the technologized and the nontechnologized, the social and the aesthetic, we have seen how this founding myth of modernism has emerged and persisted, and how it has served as a historical mode of apprehension. In these critical paradigms, the implied conception of the operations and effects of technology is overwhelmingly based on an understanding of the nature of technology as pure instrumentalism, as an automated and essentially nonhuman process of means and ends. As Martin Heidegger argues in "The Question of Technology" (1953), the problem with instrumentalist definitions of technology stems from a more profound dilemma: those unreflected notions of determinism and causality that commonly underlie such views. Causality, Heidegger maintains, has for so long been represented as *causa efficiens*—when the silversmith makes a chalice, he "causes" it—that it has set the standard for more or less all notions of causality.[36] This is not the place to delve into Heidegger's thoughts on causality. It should be enough to point out that a crucial component of the founding myth of modernism—the split between technologized and nontechnologized—presupposes a deterministic view where the nature and operations of technology tend to be seen in mechanistic and instrumentalist terms.[37]

Raymond Williams, in "Culture and Technology" (1983), reasons along similar lines. Problematizing debates on the cultural significance of recent communication technologies, Williams identifies a common thought pattern. This is the notion that machines and technologies are more or less autogenetic, endowed with their own agency and inertia, and hence responsible for the radical and usually negative changes that appear to follow in their tracks. Given such a deterministic conception, he suggests, it is not surprising that cultural pessimism routinely follows in technology's wake. Williams's reminder is simple yet crucial. Technologies do not grow, he stresses; they are made. The emergence of new technologies is, in other words, embedded in economic, social, and political practices from the start.

What is particularly interesting is that Williams, like Adorno before him, identifies the two constitutive faces of modernism, the two halves that do not add up to a whole: high-modernist art and mass culture. He maintains that the inability to recognize their interrelatedness is due to the cultural pessimism that tends to accompany technological determinism. Mass culture and high modernism, he urges, must be understood as two related yet distinct phenomena that respond to contradictions in the social order as a whole: "The real forces which produced both, not only in culture but in the widest areas of social, economic and political life, belonged to the dominant capitalist order in its paranational phase."[38] Apart from a few references to urbanization and the concentration of capital to metropolitan centers, however, Williams does not elaborate on the nature of these forces. Nor does he dwell on how the operations and effects of technologies in the modernist period may be theorized. Yet Williams's identification of the common features of technological determinism and its role in paradigms of modernism helps explain the dominance of the notion of autonomy and the corresponding ahistorical views of modernism.

If, then, the antitechnological bias of high-modernist art is in some sense a suppression, denial, or even a renunciation of the historical, social, and institutional conditions that brought it into being, it is easy to see why the ultimate modernist aesthetic had to be a negative theory of art. Yet to base a historical understanding of modernism on its own antitechnological bias produces a weak reading of modernism. A strong reading of modernism has to be based on a reconsideration of the relations of technology and aesthetics, technology being a fundamental, even constitutive, part of modernist culture. Before proceeding to the critical-interpretive part of this book, I want to discuss briefly a series of works that offer useful points of departure.

The Translation Model

The most radical critiques of technological determinism to have emerged in recent years are to be found in scholars such as Bruno Latour and Donna Haraway.[39] For all their theoretical differences, both maintain that, to use Gilles Deleuze's formula, "machines are social before being technical. Or rather, there is a human technology which exists before a material technology."[40] The notion that technology partakes of already existing social, political, and economic relations undermines the widespread view that ideas originate in the supposedly autonomous mind of the scientist, and then spread, more or less by themselves, because of their truth value. This idealist notion, furthermore, is often coupled with the determinist idea that machines and not humans, much less social, economic, or political needs, are thought to change culture and society. Latour designates this notion of science and technology as the diffusion model.

As an alternative, Latour offers what he calls the translation model. "No one," he writes in *Science in Action* (1987), in a sentence that sums up the basic tenet of his enterprise, "has ever observed a fact, a theory or a machine that could survive *outside* of the networks that gave birth to them" (248). The accumulation of knowledge, for Latour, is fundamentally relative to the means—technological, institutional, economic, and rhetorical—with which knowledge is pursued. This approach in turn is founded on a crucial premise: that the status of a particular scientific statement, idea, or fact, even a machine, is defined by its subsequent use. This idea, incidentally, is reminiscent of how Ludwig Wittgenstein, in *Philosophical Investigations* (1953), discusses linguistic meaning in terms of language games and forms of life. If the significance of a statement, fact, or idea, even a machine, is in the hands of future users, it follows that the distinction between content and context ultimately has to collapse. What also follows is that scientists have to mobilize resources, both human and nonhuman, in order to determine the future use of ideas and facts.

One of the most efficient ways of strategically stabilizing, qualifying, and modalizing future use of ideas and facts, Latour states, is to construct a machine (128). Once an idea, fact, or machine becomes established— that is, once it is used positively within a context where the so-called actants have agreed upon the criteria for defining an acceptable use—it gradually turns into a so-called black box. The black box, then, is a way of aggregating alliances. In other words, it materializes them. The black box is thus a *translation* of the alliances it gathers together. "We may on the one hand draw its *sociogram,* and on the other its *technogram,*" Latour suggests. "Every piece of information you obtain on one system is also information on the other" (138). The scientific process is a continuous negotiation between the sociogram and the technogram; and the black box is the necessary passage point that connects the one to the other. To read the technogrammatic configuration is therefore to read the sociogrammatic one, and vice versa. This does not mean that the one level is reducible to the other, only that it is translatable *onto* the other.

For Latour, then, science and technology is a potentially unstable "network," always implicated in an open and multileveled system of social, economic, political, ideological, and technological practices. "The word network indicates that resources are concentrated in a few places—the knots and the nodes—which are connected with one another—the links and the mesh: these connections transform the scattered resources into a net that may seem to extend everywhere" (180). The network metaphor signals that the heterogeneous whole can never be grasped in its totality. It is therefore all the more interesting that Latour, while discussing the notion of the network, evokes yet another metaphor, this time a technological one, so as to represent its elusive yet all-embracing character: the telephone network. "Telephone lines [. . .] are minute and fragile, so minute that they are invisible on a map and so fragile that each may be easily cut; nevertheless, the telephone network 'covers' the whole world" (180). The metaphor illustrates the fact that a small number of technoscientific actants have power over large territories or large populations. At the same time, the image of the telephone network bespeaks the impossibility of conceiving of the totality. Just like any other model, the sociological rendering of the whole will therefore always be a conceptual construct, a model of a totality that forever escapes the observer.[41]

Unlike the diffusion model, the translation model offers a universe where everything is mediated and no phenomenon is more primordial than the other. Rather, all levels are interrelated yet separate from one another. The translation model, one might say, thus proposes that every level is a signifying system which is to be translated onto other levels. For this reason, the signifying systems can usefully be seen as allegories. In fact, Latour's symmetrical mode of reading may be grasped as an essentially allegorical approach. The benefits should be obvious, for in contrast to the diffusion model, the translation model articulates the multiplicity and complexity of mediation, and this is also why Latour promotes the notion of network rather than a "social" explanation of science and technology.

In our context, this nonfoundationalist account of science and technology has several advantages. In Latour, the terms "mediation" and "translation" operate critically; they serve to deconstruct standard oppositions between, for example, content and context, machines and ideas, abstract and empirical, and human and nonhuman resources.[42] At the same time, the terms "mediation" and "translation" form an integral part of Latour's alternative conception of the technoscientific process. According to this model, machines are unstable entities, already embedded in social and institutional practices, and for this reason, machines figure as translations of the elusive yet structured social, institutional, and economic relations that partake in their constitution. Latour thus dissolves the common notion of the machine as an ever-pregiven entity. And if, as he proposes, a machine is an aggregation of social interaction, its human users may usefully be grasped as part of one and the same human-machine continuum. As I will suggest in Chapter 2, *The Magic Mountain* offers an excellent example of how sociogrammatic and technogrammatic configurations coincide in the novel's major black box—the x-ray apparatus.

Literary Modernism and Technology

In the last few decades, a number of cultural studies have placed technological change at the center of modernist art and culture, thereby calling into question a critical commonplace in traditional modernist historiography. A pioneering work of cultural history, Stephen Kern's *The Culture of Time and Space, 1880–1918* (1983) offers a wide-ranging discussion of the entire spectrum of technological inventions and scientific discoveries that emerge in the period. Scanning modernist culture from music and literature to the visual arts, including the historical avant-garde movements, Kern focuses on how the experience of time and space changes in a period that witnesses the emergence of Greenwich Mean Time as well as elevators. In this way, he challenges ahistoricist and essentialist accounts of modernist cultural practices. Yet although Kern usefully insists on the need to contextualize aesthetic modernism, his survey of technological change tends to mobilize individual works as mere illustrations. His brief discussions of cultural phenomena are limited to the level of motifs and subject matter. Moreover, Kern's study aims primarily to establish the parallels, analogies, and similarities between technoscientific culture on the one hand and modernism on the other, all subsumed under the ostensibly transcendental categories of time and space. According to this model, modernist culture testifies to technological change either by echoing it or by contesting it. Proust's notion of heterogeneous, private, and bodily time, for example, is simply posited as a reaction to so-called homogeneous time. In construing aesthetic modernism as the essential reflection of, or reaction to, technological transformation, Kern's study falls short of addressing the problem of mediation.

Marshall Berman's *All That Is Solid Melts into Air* (1982) offers a thoroughly materialist perspective on modernist art and culture.[43] Arguing that modernism can be defined only in relation to modernization, Berman suggests that it should be understood as a dialectical reaction to the spread of industries, technology, and urbanization. More specifically, modernism is constituted by the "visions and ideas" through which men and women attempt to become agents of modernization. Thus defined as a mode of historical consciousness, modernism is as old as modernization itself. In Berman's view, therefore, all thinkers from Goethe to Foucault are modernists insofar as they ponder the modern condition. Twentieth-century modernism, however, is marked by the desire to escape the inherent contradictions of the modernization process. Modernization is simply seen as either positive or negative. Thus the one group of twentieth-century modernists, emblematized by the futurists, blindly affirms all aspects of modernization, including technology; the other group rejects modernization just as blindly. Berman's examples of this latter group are Max Weber, Ortega y Gasset, T. S. Eliot, Oswald Spengler, and Allen Tate. In short, having lost the capacity to grasp the contradictions of modernization, twentieth-century modernists are unable to appreciate the extent to which this process is both liberating and constraining. But that Berman reaches this conclusion testifies to the fact that he, too, is caught in the dualisms through which modernism is typically conceived, and this, in turn, stems from the fact that he tends to posit modernism as an idea—a vision, a *Weltanschauung*.[44] Furthermore, what is peculiarly absent in Berman's study is a critical engagement with those texts traditionally referred to as modernist, save for Baudelaire's poetry. As I will show, however, once certain stylistic, rhetorical, and other formal aspects of high-modernist texts are subjected to analysis, it becomes clear that they register a variety of contradictions intrinsic to what Berman designates as the modernization process.

Friedrich Kittler's *Discourse Networks* (1985) marks a new and theoretically sustained departure in the attempt to articulate the relations of technology and modernist aesthetics.[45] Where Kern and Berman discuss the relations of technological change to notions of time and space, Kittler's focus is more limited; his concern is with the operations of media technology, more specifically with what he calls inscription technologies—notably, the alphabet, the typewriter, phonography, and cinematography. This allows Kittler to tell an alternative literary history, one that begins, not with the text or the word, but with the materiality of media. In his view, the materiality of media precedes processes of signification and, by implication, the institution of literature. Matter, in short, is prior to spirit. Kittler's aim is therefore to develop a "discourse analysis" capable of making visible the material conditions of possibility for meaning to occur in the first place.

Steeped in a Foucauldian framework, particularly the rhetoric of epistemic shifts as practiced in *The Order of Things* (1966), Kittler's inquiry traces two paradigm shifts in this predominantly German history of letters. The first break, taking place around 1800, occurs with so-called universal alphabetization. The second shift happens around 1900, initiated by the advent of new technologies for storing sense data, notably phonography and cinematography. With the emergence of this new "Aufschreibesystem," the previous discourse system is radically and irrevocably relativized. The power of the written word reigned supreme until 1900, Kittler argues, but at the turn of the century it was severed from the animating spirit that used to support its ever-pregiven meanings. The letter emerges as a material shell, hence potentially empty, now exposed in all its exteriority and sensuality. Meanwhile, the book loses its monopoly on storing sense data, henceforth forced to yield to two newcomers: phonography and cinematography. This, simply put, is the thrust of Kittler's genealogy of German letters, a rich cultural history whose trajectory sheds new light on Fichte's views on meaning and Edison's aesthetics of hearing, as well as Freud's conception of mnemonic techniques and Rilke's experience of sonic texture. Still, the textual analyses of modernist works in *Discourse Networks* are surprisingly few and also cursory, now based in diary entries, now in interviews, now in literary texts. This shortcoming is probably linked to the fact that Kittler, like Foucault before him, rejects (a certain conception of) hermeneutics as well as (a certain conception of) dialectics. For a vastly ambitious project like Kittler's to be fully convincing, however, a careful and sustained textual analysis has to be placed center stage.

Even more important, however, *Discourse Networks* fails to account for a number of theoretical questions. How, for example, is an effective theory of change possible from the point of view of discourse analysis? What is the relation between a social formation and a discursive formation? Furthermore, Kittler's account of the discursive a priori specific to the twentieth century gives historical priority to technologies of inscription; he implicitly ascribes historical agency to the typewriter, the phonograph, the gramophone, and the cinematograph.[46] At the same time, Kittler's primary concern is less with technology and more with a certain notion of writing. The phenomena under discussion—the typewriter, cinematography, and phonography—are all posited as "media" or, alternatively, "inscription technologies," which, it will be noted, is already to theorize them. Cinematography, after all, may equally well be approached as a technology of perception, or as a technology of reproduction, or as a communication technology, or as a visual technology; each of these perspectives naturally prompts different histories. What is at issue in Kittler is not aesthetics, not perception, not consumption, not exchange, but rather writing, that is, the material scene of writing, including the writing body, whether human or nonhuman. Indeed, the notion of writing—*schreiben* and *aufschreiben*—is the axis around which his inquiry pivots.

This, then, provokes my final point. In Kittler, it is the discourse system that serves as a mediating instance between processes of modernization on the one hand and literary practice on the other. In Kern, as we have seen, what mediates between modernization and modernism is the experience of time and space, while in Berman, finally, it is the notion of "vision and ideas" or "world-view." What these approaches have in common is a tripartite scheme where the literary artifact is readily seen as a reflection of modernization or, conversely, as a reaction to it. While such schemes are often heuristically necessary, we need to develop a more complex and heterogeneous notion of mediation.

Tropes of Literary-Historical Imagination

How, then, does the inner logic of the individual work register technological change? And how is such textual sedimentation to be conceived of and articulated? As my discussion of the cultural studies of Kern, Berman, and Kittler has suggested, these complex questions are theoretical, historical, and representational all at once. In more properly literary-critical studies of technology and modernism, two major tendencies can be distinguished, and I will consider them, briefly, under the headings of allegory and symbol.[47] As we saw in the context of Latour's translation model, an allegorically operating critical discourse stresses likeness and continuity as much as difference and discontinuity. Furthermore, it works to indicate how multiple causalities are at work simultaneously.

A symbolically operating critical discourse, by contrast, emphasizes likeness, identity, and organic unity. It aims primarily to establish analogies and similarity. As a result, literary discourse is all too readily posited as the essential *expression* of a prior or underlying technoscientific logic, force, or mechanism. Hugh Kenner's *The Mechanic Muse* (1987), a study of Anglo-American modernism and its relations to technology, is representative of this approach. In Kenner, metaphorics comes as a solution to the theoretical and representational problems that all studies in this interdisciplinary field have to deal with implicitly or explicitly. In his discussion of Eliot's *The Waste Land*, for example, Kenner proposes that telephony may usefully be related to the thematics of the poem. He concludes that "*The Waste Land* is, so to speak, a telephone poem, its multiple voices referrable to a massive short-circuit at the central exchange."[48] Kenner here invents a simile. Using the language of telephony to rewrite the language of the poem, including its specific workings and meanings, he collapses two distinct signifying systems into one—without accounting for the mediating instances that both separate and determine them. In other words, Kenner's rhetoric operates symbolically. The one level is effectively seen as an expression of the other, for the simile serves to suggest likeness, even identity, between the two distinct levels whose semiautonomy are thereby conflated. This is not to say that it is possible to manage without tropes, only that it is essential to reflect on the underlying assumptions of Kenner's penchant for metaphor, since such modes of articulation, to my mind, stand in for a more rigorous, historically reflexive attempt at describing the relations of technology and art.

Yet such a rhetoric of likeness and congruity is a constant temptation in studies of technology and culture. Consider, for example, George Landow's *Hypertext* (1992).[49] Seeking to demonstrate that recent critical theory prefigures and inscribes phenomena such as hypertext, Landow can be said to show that the particular *idioms* that are used to describe the nature of cyberspace and its cultural and social consequences, especially as they pertain to literary scholarship, coincide with the *idioms* of recent critical theory. For such an argument to become effective, however, its premises must include a reflection upon the conceptual, social, and ideological conditions that enable these distinct languages and their separate histories of constitution.

Apart from the simile and other symbolically operating tropes, another, highly specific rhetorical figure resurfaces in a number of critical texts, particularly in studies that claim that a specific material condition has bearing upon consciousness or subjectivity. The figure assumes the form of "A relates to B as C relates to D." Thus Walter Benjamin, in his famous essay "The Work of Art in the Age of Mechanical Reproduction," writes that "magician and surgeon compare to painter and cameraman."[50] In *The Mechanic Muse,* Kenner states: "*Ulysses:* a book as remote from old ways to tell a story as the linotype is from the old way to handle type."[51] Similarly, Wolfgang Schivelbusch maintains: "The railroad related to the coach and horses as the modern mass army to the medieval army of knights (and as manufacture and industry do to craftsmanship)."[52] Finally, Katherine Hayles suggests, though in a different context, that "cybernetics is connected to virtual reality technologies in much the same way as Cartesian space is connected to contemporary mapmaking."[53] What is at issue here is the problem of how to represent the hypothesis that one realm of social life, or system of signification, bears a relation to another, and how to represent that connection. The rhetorical figure of A:B::C:D can then be seen as a tropological solution to a historiographic problem. Aristotle, in *Poetics,* identified this figure as an analogy or proportion, on the basis of which one might construct a metaphor: "Analogy or proportion is when the second term is to the first as the fourth to the third. We may then use the fourth to the second, or the second for the fourth."[54]

Aristotle spoke primarily of tragedy and literary language; analogy, in his view, offered embellishment and aesthetic gratification. In our context, the figure serves as a trope of historical imagination, providing a sense of similarity and proportion by positing a structural analogical relation between two separate realms. But while the figure states that a parallelism exists, it naturally cannot explain the relationship as such. The analogical move, whose main function is to establish similarity, must therefore be complemented with differential elements: different effects, different configurations, different environments. In this context, it is instructive to scrutinize Benjamin's use of tropology, more specifically, his use of allegory as a critical-interpretive device.

Allegory as Mediation

One of the most sophisticated efforts to address the intersections of modernist culture and technology is found in the works of Benjamin, both as manifested in *The Arcades Project* and in his essays on modernist artists and writers such as Baudelaire, Breton, Atget, Proust, Kafka, and

Brecht. As Susan Buck-Morss has demonstrated, the "relationship be-tween art and technology is a central theme in [*The Arcades Project*]."[55] Given the unfinished nature of Benjamin's great project, however, it is only in the essays that we may distinguish more clearly how he conceived of the relationship between art and technology. In a number of articles, most of which were written during the 1930s, Benjamin commented on Baudelaire's work, in particular *Flowers of Evil*.[56] The language of the French poet, in Benjamin's view, accommodated unprecedented forms of historical experience, ultimately registering changes in the economic mode of production. In attempting to trace the remote yet palpable ef-fects of industrial capitalism in *Flowers of Evil*, Benjamin deliberately used a seemingly fragmentary and open-ended mode of critical writing, con-sistently deploying images rather than concepts. These features were all part of Benjamin's idiosyncratic theory of historiography, one that also served as a cultural theory of modernity.[57]

Speaking of the method of *The Arcades Project*, Benjamin once re-marked that one of its most important and most difficult tasks was to "demonstrate that the materialistic presentation of history is imagistic," and far more radically so than traditional historiography.[58] In this spirit, he afforded images a central role in "Some Motifs in Baudelaire" (1939). Indeed, a crucial component of the formal structure of the essay is the concrete image and the intricate ways in which it variously enters into analogical and differential relationships with other images. By exploring this structure, I want to suggest that Benjamin stages an allegorical mode of representation that ultimately serves as a model for mapping media-tion and overdetermination.[59]

Images are located at the heart of the essay—eyes, fingertips, hand movements, glances, heads. In fact, entire scenes are presented to the reader: Baudelaire thinking about his readers, the worker at the assembly line, the gambler at the roulette table, the crowd and the flâneur, the ar-cade and the boulevard, as well as a host of technological innovations from the camera to the match. Yet this is only part of Benjamin's method, as these images and scenarios also work to create tension-filled juxtaposi-tions. As a constellation becomes visible, a tentative thematics evolves. When a new image is added, what first appeared marginal emerges as a central preoccupation, and vice versa. In this way, what initially appeared as concrete and particular turns into a figuration—an allegorization—of something else. Ultimately, allegorization becomes a device for construct-ing mediation.

The central question in "On Some Motifs in Baudelaire" is the status of poetry in the age of technology and capitalism.[60] At the same time, the essay implicitly speaks of the difficulties facing historical representation in the modern. Read alongside "The Storyteller" (1936) and "The Work of Art in the Age of Mechanical Reproduction" (1936), the Baudelaire essay suggests that Benjamin, as a cultural historian, is confronted with much the same problem as the storyteller in the modern period. History, encapsulated by technological change, has exploded the conditions of existence of storytelling. Different modes of representation are called for, and this is where allegory enters the picture.

The Baudelaire essay begins as a reading of how the predilection for sensuous experience is configured in three writers. Baudelaire's *correspondances*, Proust's madeleine, and Bergson's *durée*, Benjamin suggests, could come into being only when "genuine" forms of experience had become subject to "disintegration." In order to explore what triggered the desire for "crisis-proof" (*krisensicher*) experience in the first place, Benjamin gathers reflections on the organization of urban space, the crowds in the city, the worker at the assembly line, innovations such as the camera and the telephone, and other typically modern motifs. But if the question at issue is the disintegration of experience, why is it necessary to deal with Baudelaire in the first place? Because, as Jürgen Habermas has underscored, the Benjaminian theory of art is a theory of experience.[61] Baudelaire's poetry, then, is an index of preeminently modern modes of experience. In Benjamin's view, *Flowers of Evil* testifies to the price that had to be paid for the advent of the modern: disintegrated experience is substituted for genuine experience. Baudelaire thus represents a historical threshold, his poems making visible how bourgeois forms of social life intersect with those of emergent mass society. As Benjamin takes a closer look at Baudelaire, he observes a twofold and typically modernist impulse: on the one hand, the poet was a traumatophile, in pursuit of the absolutely new and the not-yet-domesticated; on the other hand, he dreamed of a synthetic and crisis-proof experience—in the form of *les correspondances*.

The essay opens with a concrete historical situation: "Baudelaire," Benjamin states, "wrote a book which from the very beginning had little prospect of becoming an immediate popular success" (109). The reasons may be glimpsed in the very fabric of *Flowers of Evil*, Benjamin suggests, as the poems themselves reflect the historical circumstances that help explain the relative indifference of the contemporary audience. Reflecting on Baudelaire's motifs, Benjamin dwells on them until each and every one unfolds into its historical specificity. There is no story to be told, however, no linear argument that accumulates before the reader's eyes. Rather, the essay is a series of beginnings, which turn out to be returns: Benjamin tirelessly approaches the central problem—the disintegration of experience—by way of tortuous detours, expecting the particular to open up the path to the secret whole.

As Benjamin turns to contemporary philosophy for reflections on why forms of experience may have changed, he finds that there, too, these changes have left their mark. This is where Bergson enters the argument, both as a symptom and as a relay: a symptom because Benjamin thinks of Bergson's philosophy of intuition as a negative reflection of the historical circumstances that brought it into being; a relay because it provides a link to Baudelaire. Bergson's idea was, in short, that *la durée*—the duration of pure memory—would be a way of mastering the instrumentality of everyday life, thus freeing the subject from what he called the rhythm of necessity. In Benjamin's view, however, Bergson's theory of pure memory creates a "spontaneous after-image," and this after-image is the historical specificity that Bergson failed to recognize. In this essentially negative way, Bergson furnishes a clue to the structure of experience with which Baudelaire was confronted and which he attempted to accommodate. This, then, is the first constellation. Bergson's theory figures as a negative imprint of poetic structures characteristic of Baudelaire. At the same time, Bergson also figures as a parallel, since the wish for "crisis-proof" experience links Baudelaire's notion of *correspondances* to Bergson's philosophy of intuition.

The next constellation is Bergson and Proust. Influenced by Bergson's notion of pure memory, Proust elaborated a distinction between voluntary and involuntary memory. The quintessential moment of Proustian *mémoire involontaire* is, of course, when the narrator recaptures things past thanks to the madeleine. In Benjamin's view, this suggests that Proust, too, pursues a form of crisis-proof experience. The sensory impression, in bringing back past experience in all its uniqueness, temporalizes the self by way of the individual body, thereby warranting a sense of identity. Benjamin's point is that the seemingly timeless desire for such experience—whether in Baudelaire, Proust, or Bergson—confirms the historical circumstances it seeks to transcend. The desire for sensory irreducibility suggests not only that the world of the living is increasingly ruled by an instrumental logic, but also that modes of sensory perception have likewise become subject to a logic of means and ends.

What, then, are the relations between Baudelaire's poetry, spleen, *les correspondances*, Bergson's theory of memory, the flâneur, Proust's madeleine episode? And if, as Benjamin maintains, these phenomena are interrelated, what is the mode of representation proper to their mapping? In order to show how Benjamin attempts to solve such problems, I shall focus on another constellation, where the concrete image is turned into allegory. Benjamin evokes Marx's analysis of the worker at the assembly line—in the capitalist mode of production, Marx argued, it is the instruments of labor that employ the workman, not the other way round—and then suggests that technological innovations give rise to a an all-pervasive logic of mechanization, one that pertains not only to the worker's adaptation to the machine but to social relations at large. In fact, forms of consciousness, too, are affected, as technology brings about what Benjamin calls "shock experience." Shock is a specifically modern condition, from the small-scale ones triggered by the match, the telephone, and the camera, to those large-scale concentrations of perceptual stimuli inherent in urban space. Hence the key statement in the essay: "Technology has subjected the human sensorium to a complex kind of training" (132). The more consciousness has to engage in screening off the stimuli of modern life so as to ward off trauma, Benjamin argues, the less likely it is that these sense impressions will enter experience (*Erfahrung*); this, too, is a specifically modern predicament.

The passage about the worker serves the purpose of demonstrating how in modern society older life patterns, for example, apprenticeship and storytelling, are forced to subordinate themselves to a logic of means and ends. Characteristically, Benjamin adds that the "shock experience which the passer-by has in the crowd corresponds to what the worker 'experiences' at his machine" (134). This structural analogy serves to epitomize the changes of experiential modes in general. At the same time, however, by turning the worker at the assembly line into allegory, the analogical move erases the specificity of the conditions of factory work, and the image of the worker comes to signify the shocks of modern life and the related atrophy of experience in general.

Two kinds of mediating relations are hinted at in this passage. On the one hand, modes of experience are dependent upon the position of the individual in the economic system. On the other hand, the mode of production dominated by machine-powered technology brings about the same psychological effects for all, regardless of class position or place in the production process. What is implied here is that multiple causalities are at work simultaneously. Put differently, the image of the worker at the assembly line serves to represent overdetermination. Rephrased in this way, this and the other allegorical images justify Benjamin's assemblage of seemingly unrelated narrative fragments.

The next constellation, proposing yet another allegory, amplifies Benjamin's image of the mechanized worker. Baudelaire, he writes, was "captivated by a process whereby the reflecting mechanism which the machine sets off in the workman can be studied closely, as in a mirror, in the idler" (134). Benjamin thinks of the gambler at the roulette table, another motif in Baudelaire. The gambler emerges as a parallel of the worker at the assembly line; neither can make use of experience, both being forced to adapt to mechanization, repetition, and "shock."

In this way, Benjamin's constellations trigger a continuous dialogue between sameness and difference. First, Bergson provides both a negative allegory (his denial of history) and a parallelism or analogy (*la durée*); second, the experience of the passer-by in the crowd parallels that of the worker at the machine; third, the chopped-up time at the assembly line corresponds to the kind of discontinuous temporality that Proust traced in Baudelaire; fourth, this experience of time is turned into negative allegory by *les correspondances,* where temporality, pure and continuous, stands out from historical time, and so on. Benjamin's repertory of allegorized images thus suggests how Baudelaire's conception of *Flowers of Evil* is overdetermined by an array of historical phenomena. Marked by the disintegration of experience, Baudelaire's motifs are inscribed by a host of historical conditions, minor and major, vertical and horizontal, such as the invention of the match, big-city traffic, urban crowds, production technologies, commodity fetishism, and so forth. Yet nothing is seen as more primordial than the other. The whole remains secret, evoked negatively by way of allegorization.

In placing allegory center stage, Benjamin manages to show that multiple causalities are at work simultaneously. He also manages to show that in Baudelaire, the crisis in sense perception translates onto the crisis in artistic reproduction, and vice versa. Allegorization offers the possibility of thinking two things at the same time, thus enabling a dialectical mode of representation. Adorno once spoke of the essay as form, arguing that it "does not glorify concern with the original as more primordial than concern with what is mediated, because for it primordiality is itself an object of reflection, something negative."[62] In Adorno's view, Benjamin's essays were exemplary. Allegorization, as I have proposed, was Benjamin's primary theoretical tool. It was intimately linked to the unspoken acknowledgment that Benjamin, as a cultural historian, was part of the same historical process as the phenomena upon which he reflected. This, I believe, is the distinguishing mark of his critical method.

To be sure, the historiographic model that Benjamin puts to use in this and other essays is neither systematic nor unproblematical. At times, Benjamin's engagement with Baudelaire's writings is superficial; he also fails to introduce distinctions between various kinds of technology. Nonetheless, important aspects of Benjamin's method may be appropriated, in particular the use of allegory as a mode of representing mediation. Once we look upon the bewildering variety of technologies, machines, and apparatuses that emerged during the second half of the nineteenth century as allegories of the ways in which essentially social relations are reconfigured, then we begin to discern how the topography of aesthetic modernism might be rethought. Those cultural practices that we know as modernism might then be seen as forms of "crisis management," not exterior to but already immersed in those conditions—technological, economic, social—it sought to go beyond.

2 Novel Visions and the Crisis of Culture

The Cultivation of the Interior in

The Magic Mountain

Who can still believe in the opacity of bodies, since our sharpened and multiplied sensitiveness has already penetrated the obscure manifestations of the medium? Why should we forget in our creations the doubled power of our sight capable of giving results analogous to those of the X-rays?

UMBERTO BOCCIONI ET AL., "FUTURIST PAINTING: TECHNICAL MANIFESTO," 1910

In the mid-nineteenth century, the artist's eye begins to claim sovereignty with unprecedented energy.[1] Indeed, in the period that sees the emergence of mechanical devices for reproducing visual phenomena—from the camera to the X-ray machine—the artist's eye increasingly claims autonomy from the habits of artistic perception inherited from the tradition, and above all from scenarios of rational, instrumental, or generalizing knowledge.[2] To move from Claude Monet's sunstruck haystacks to Man Ray's somber rayographs, from Eugène Atget's deserted Parisian streets to Umberto Boccioni's speed-infused sculptures, is to bear witness to the stunning diversity of the modernist conquest of the visual. No longer located in the ideality of the sense of sight, visual perception is now grounded in the bodily being of the individual in all its ostensible immediacy. Vision is celebrated for vision's sake. *Aisthesis* is invented anew.[3]

In *Techniques of the Observer*, Jonathan Crary argues that the model of vision that modernism deploys, whether implicitly or explicitly, takes as its point of departure the immanence of the individual body, not the Cartesian, noncorporeal, and transcendental model of vision.[4] Loosely following Crary, one might say that *theoria* is increasingly associated with vision machines, such as Etienne-Jules Marey's chronophotographic camera, Wilhelm Conrad Röntgen's X-ray machine, and other optical technologies that similarly explore and articulate physiological domains otherwise inaccessible to the human eye. In ancient Greek *theoria* means a looking at, viewing, contemplation, speculation, theory, but also a sight, a spectacle.[5]

If, then, the artist's eye claims independence, self-sufficiency, and irreducibility, it is a utopian gesture that must be understood as marked by, among other things, the technoscientific discourses that helped produce the new matrices of perception. J.M.W. Turner once said, "My business is to draw what I see, and not what I know is there."[6] Seeing vis-à-vis knowing: this particular division of labor between the aesthetic and the epistemic, the sensuous and the systemic, *aisthesis* and *theoria,* is intrinsic to nineteenth-century modernity in general and to the emergence of various photographic technologies in particular, and it should be seen as coextensive with a radical epistemological crisis.

Few modernist novels dramatize this crisis as effectively and obsessively as Thomas Mann's encyclopedic novel *Der Zauberberg* (1924), translated as *The Magic Mountain.* Thematizing the workings of the eye and the limits of what it may see and know, all the way from social scanning and the medical gaze to the mechanical eye, *The Magic Mountain* emerges as an inquiry into the relativization of the epistemic mandates of human vision. The narrative connects this epistemological crisis to the theme of *Bildung* and, by implication, subjectivity. As we shall see, Mann's novel details the hero's first encounter with an array of new technologies, including an X-ray machine, a cinema show, and a gramophone; each can be seen as a signpost along the protagonist's long and cumbersome road of education. Yet although the thematics of technology is richly orchestrated in *The Magic Mountain,* scholars and critics have failed to recognize its importance, save for a passing mention in a few studies.[7] Nor has the central issue of vision and visuality received critical treatment. In this chapter, therefore, I want to propose a new interpretive framework with which to approach Mann's best-selling novel. By tracing how the experience of the machine is represented, both on the level of enunciation and on the level of form, I want to argue that *The Magic Mountain* is inscribed in the debates concerning the relations of *Kultur* and technology that were so prevalent in Weimar Germany. I also want to propose that the novel stages the problem of *Bildung* and modes of subjectivity in terms of visually signifying systems. Indeed, Mann's modernist aesthetic must be thought together with the emergence of visual technologies that made available new optical spaces and new realms of knowing, as well as the technologies of reproduction that enabled the mass production of cultural artifacts.

An Exemplary Modernist Novel

The story is well known. Hans Castorp, a young engineer, goes to visit his cousin Joachim at a sanatorium in the Swiss Alps. The year is 1907. A perfectly healthy young man, he intends to stay for a mere three weeks. He stays at the sanatorium for seven years, however, not because he has to, but because he wants to, even after his cousin's premature death. Harassed by a slight fever, the engineer becomes an intellectual, a learned man, voraciously studying anatomy, physiology, biology, botany, astronomy, and other subjects in order to explore what he calls "the nature of man," including the meaning and purpose of life. He also falls in love, desperately and hopelessly, with a Russian woman, Clavdia Chauchat. After seven years, Castorp, now thirty years old, decides to enroll in the army and finally descends from his chosen Olympus. The narrative comes to a cinematic end as our hero, bayonet in hand, disappears from sight, staggering along in the steady rain while he sings a little song. It is as though the camera is slowly pulled back, as the protagonist walks into the unknown and the credits roll before the spectator's baffled eyes. The story thus comes to a close well before the denouement of the classic bildungsroman has been set in place. The point is, of course, that the traditional kind of closure was structurally impossible. A deliberate parody of the classic bildungsroman, *The Magic Mountain* everywhere feeds on a promise that it can never fulfill: the author saw to it that his unassuming young protagonist, unlike the hero of the traditional bildungsroman, lacked a past that could claim his destiny.

Mann started work on the novel in 1912, a couple of years before the onset of the Great War, and finished it only in 1924.[8] As Hans Levander has pointed out, *The Magic Mountain* unexpectedly became a best-seller. In 1928, more than one hundred thousand copies had been printed.[9] Intended as a novella, *The Magic Mountain* was meant as a humorous and satiric counterpoint to *Death in Venice* (1912), but it swelled into a full-scale novel. The result was an encyclopedic book on a par with other striking literary works published in the period, most prominently Proust's *Remembrance of Things Past* (1913–27), Joyce's *Ulysses* (1922), Robert Musil's *The Man without Qualities* (1930–32), and Hermann Broch's *Sleepwalkers* (1930–32). Italo Calvino has designated them as "open" encyclopedias: "What tends to emerge from the great novels of the twentieth century is the idea of the *open* encyclopedia, an adjective that certainly contradicts the noun *encyclopedia*, which etymologically implies an attempt to exhaust knowledge of the world by enclosing it in a circle."[10]

The more the modernist novels attempted to exhaust the world, to provide that all-encompassing representation of the human realm in its elusive totality, with its mind-boggling minutiae, the more they had to bear witness to the impossibility of the project itself. Calvino's suggestion is particularly relevant in our context, since *The Magic Mountain*, more than anything else, is a novel about knowledge—its conditions, its processes, its consequences.

The classic bildungsroman, it has sometimes been claimed, is the story of how the male bourgeois subject is constituted, the ultimate aim being harmony between the individual and the social. Castorp's trajectory, of course, deviates from the classical pattern. His *Bildung*, for one, is made possible not by means of traveling but by a closed, disciplinary space—the sanatorium. Like Frédéric Moreau in Gustave Flaubert's *Sentimental Education*, Castorp thus secures for himself a space of social freedom. As Pierre Bourdieu remarks in his analysis of Moreau's social context: "The tendency of the patrimony (and hence of the entire social structure) to persevere in itself can only be realized if the inheritance inherits the heir, if the patrimony manages to appropriate for itself possessors both disposed and apt to enter into a relation of reciprocal appropriation."[11]

Symbolically enough, in Castorp's case, just as in Moreau's, the implicit claims of the patrimony are cut short. Indeed, because Castorp becomes an orphan at an early age, the patriarchal descent of the family is broken, a family of prosperous merchants and senators whose proud lineage goes back at least to 1650. Given that the novel so emphatically denies its protagonist a paternal past that may appropriate his future being, it is interesting to observe that the author himself makes sure that he possesses a past that may claim his writerly destiny. By inscribing the narrative in the literary paradigm fathered by Goethe, Mann makes sure to construct a generic past for himself and, therefore, also a future. To paraphrase Bourdieu: Goethe inherited Mann.

With *The Magic Mountain* Mann inscribed himself in a number of mighty cultural traditions, not only that of the bildungsroman.[12] For instance, in choosing the title of the novel—*Der Zauberberg*—Mann enters into dialogue with Nietzsche.[13] In *The Birth of Tragedy,* Nietzsche speaks of ancient Greece and its Olympian magic mountain, where the Greeks located their gods.[14] Mann, however, transforms this heavenly sphere into a space that is part bourgeois interior, part high-tech medical center, and populates it with an international crowd of good-natured sufferers.[15] Further, by foregrounding the linkages between disease, the imminence of death, and spiritual expansion, Mann inserts his narrative in a specifically romanticist tradition.[16] According to this legacy, health is associated with stupidity, mindlessness, and vulgar aspirations. The slogan "Syphilisation is civilisation" is not an unfitting rubric for this long-standing philosophical interest in the connections between genius and disease.[17] Also, because Mann includes mythological and religious substructures in the text, it resonates with references to Greek and Roman mythology as well as Christian symbolism.[18] In this way, Mann's narrative method readily appears akin to that of other modernist writers who devised techniques for creating multiple layers of meaning in the text and endowing individual existence with universal significance; it should be enough to mention Eliot's *The Waste Land* and Joyce's *Ulysses.* Finally, Mann's narrative joins the ranks of the experimental modernist novel because of its pronounced desire to delve into the enigma of temporality and unfold into a "time romance," on the level of reading as well as on the level of plot.[19] These critical traditions and their inherent languages have also couched the vast majority of interpretations of *The Magic Mountain.*

In this chapter, I want to claim another family likeness for the novel. This is not a vertical kinship, as it were, but a horizontal one: first, a kinship with contemporaneous issues, such as the modernist preoccupation with regimes of sight and visual machinery, resonating in futurist manifestos and Bauhaus programs ("Penetration of the body with light is one of the greatest visual experiences," as László Moholy-Nagy once remarked); and second, a kinship with the charged discourse concerning the relations of *Kultur* and technology that, as Jeffrey Herf has shown, was so prominent at the time.[20]

In the classic bildungsroman, time is a mere medium. It marks out the distance traveled and propels the hero forward until, eventually, he returns home and time freezes into an eternal now. Franco Moretti, arguing along these lines, observes that in the classical version of the genre, the ending and the aim of narration are brought together: "The happy ending, in its highest form, is not a 'dubious success,' but this triumph of meaning over time."[21] In *The Magic Mountain,* the tables are turned. Time is no longer a mere function of space. Indeed, time now triumphs over meaning and offers itself as the ultimate existential question: What is the meaning of time? Its nature? Its essence? Symptomatically, space is no longer a question of origins and hence also not of *telos;* at length, therefore, the reader comes to contemplate the peculiar relation between, on the one hand, spatial stasis, that is, the sanatorium, and, on the other, a narrative designated by Mann as a "time romance," which sets out to *perform* the temporal enigma itself. Stripped of its spatial correlation, temporality becomes an abstraction and is represented as such.

In the era of Goethe, according to Moretti, the bildungsroman enacts a symbolic resolution of a specifically bourgeois dilemma: How to manage the incompatibility of individual freedom and social integration? The primary means for achieving an organic unity between these two clashing desires is the symbol. This is why the bildungsroman does not accommodate the idea of interpretation per se: "To do so would be to recognize that an alterity continues to exist between the subject and his world, and that it has established its own *culture:* and this must not be. That clash, that social strife which, on the cognitive plane, the act of interpretation keeps open and alive, is sealed by the beautiful harmony of the symbol. Or in other words: meaning, in the classical *Bildungsroman,* has its price. And this price is freedom."[22]

Castorp, by contrast, has all the freedom in the world. He is free to choose, free to leave, free to fall in love, free to spend seven years in a sanatorium, free to reinvent himself. More important, however, is this. Just as temporality disengages itself from space, just as time acquires depth and turns into an abstract problem in *The Magic Mountain,* so the reader witnesses yet another dialectical leap: the problem of interpretation, along with that of allegory, alterity, and difference, reintroduces itself. It comes to the very fore of the narrative, as though we were witnessing the return of the repressed; and more often than not the problem of interpretation is linked to the image of technology.

Visual Regimes and the Formation of the Hero

The International Sanatorium Berghof is a place where well-to-do sufferers of numerous nationalities, mostly European, take their cure. A commonly held view is that the image of the sanatorium represents the decline of humanity in general and of Western civilization in particular.[23] By the same token, the sanatorium affords an existential inflection of the significance of the narrative as a whole, evoking timeless human concerns, such as disease, death, and suffering. But what if one conceives of the sanatorium as a historically specific image and not merely as a symbol? Indeed, the sanatorium, a restricted space of medicine and high technology, is a vital part of the logic of the plot, pertaining in significant ways to the central problem, as stated by the novel itself: What is life? What is man? What is the body? Indeed, in *The Magic Mountain* death and temporality are part of a cluster of notions that revolve around the image of the machine and other technological artifacts, in particular the X-ray machine.

The sanatorium is represented as a place where one engages in visual activities. One looks at other people, into optical toys, at paintings, at photographs, at X-ray plates, at the landscape, and so on. One is also looked at: by other people, by the physicians, by the X-ray machine. In fact, *The Magic Mountain* is marked by an obsession with eyes, vision, and visuality. On virtually every page, the reader finds references to sight. Eyes are variously "bloodshot," "Kirghiz-shaped," "watering," and so forth. Visual metaphors abound. For instance, Dr Behrens asks Castorp: "Well, my innocent bystander [*Zuschauer*], [. . .] what are you up to, have we found favor in your searching eyes [*Blicken*]?" (*MM* 104/*Z* 147) Furthermore, eyes are endowed with agency and impetus: "[His eyes] simply would not remain closed now, but kept fluttering open restlessly the moment he shut them" (*MM* 87/*Z* 124–25). This tendency holds not only for eyes but for body parts as well. The reader thus begins to glimpse the contours of a body that has begun to dismember itself, a body whose various parts take on a life of their own.

Generally speaking, the sanatorium harbors two major regimes of sight, both of which can be divided into subregimes. Each of these visual economies brings into play a particular division of visual labor, as ambivalent as it is stratified. On the one hand, vision emerges as *aisthesis*, that is, as a form of corporealized and individualized perception, often libidinally inflected. Vision, then, is a means of leisurely activity, pleasure, and pain (optical toys, cinema, and above all eroticism). On the other hand, vision emerges as a matter of *theoria*, that is, as a means of gathering systematic knowledge (definition, classification, typology). Or rather, theoretical vision reveals itself as a vehicle of a signifying system that maps a certain knowledge onto an object seen. This second visual regime encompasses at least three modalities: social scanning, the medical gaze, and the mechanical eye. As we shall see, these visually determined signifying systems all bear on the formation of the protagonist, mediating his experiences at the sanatorium. At the same time, however, these visual regimes are riddled with contradictions, ambivalences, and blurred boundaries. Not only do they contest and compete with one another; they also reveal themselves as subject to the vicissitudes of interpretation.

But apart from this visual division of labor, there is another division of labor at work. It has to do with perception at large. Vision is privileged, functionalized, and in some ways reified in the world rendered by *The Magic Mountain,* whereas the other senses are relegated to the margins of a universe of increasing visuality and abstraction. They wait in the wings, however, offering new opportunities for aesthetic intensities. Consider, for example, the sublime world of listening, mediated by the technologized reproduction of music, that Castorp's ears will enter at the end of his spiritual journey.

On the one hand, then, the sanatorium is a space of visual pleasure. First, and perhaps most obvious, there are the optical toys with which the patients amuse themselves in the evenings—just as in the opening pages of Proust's *Remembrance of Things Past,* where a kaleidoscope, a kinetoscope, and a magic lantern suggest an alternative world of visuality. "In the first social room," Mann writes, there were

> a few optical gadgets for [the patients'] amusement: the first, a stereoscopic viewer, through the lenses of which you stared at photographs you inserted into it—a Venetian gondolier for example, in all his bloodless and rigid substantiality; the second, a long, tubelike kaleidoscope that you put up to one eye, and by turning a little ring with one hand, you could conjure up a magical fluctuation of colorful stars and arabesques; and finally, a little rotating drum in which you placed a strip of cinematographic film and then looked through an opening on one side to watch a miller wrestle with a chimney sweep, a schoolmaster paddle a pupil, a tightrope-walker do somersaults, or a farmer and his wife dance a rustic waltz. Laying his chilled hands on his knees, Hans Castorp gazed into each of these apparatuses for a good while. (*MM* 82/*Z* 117)

The optical toys Mann describes—a stereoscope, a kaleidoscope, and a cinematographic drum, most likely a zoetrope—bring attention to the discrepancy between vision and the object seen. Like the panorama, the diorama, and the phenakistiscope, they were all invented in the first half of the nineteenth century. Their presence certainly bestows an old-fashioned aura on the patients' evening pastimes, especially since they have access to a movie theater. Yet it is peculiarly fitting that these contrivances are present in this space of medicine and optical technology, for they share crucial features with scientific devices developed in the first half of the nineteenth century for studying empirically the laws of physiological vision.

In 1859, Oliver Wendell Holmes, who invented the stereoscope and also happened to be a Harvard professor of anatomy, prophesied that a new optical era was in the making: "We are looking into stereoscopes as pretty toys, and wondering over the photograph as a charming novelty; but before another generation has passed away, it will be recognized that a new epoch in the history of human progress dates from the time when He [. . .] took a pencil of fire from the 'angel standing in the sun,' and placed it in the hands of a mortal."[24] Evoking death and finitude, Holmes locates vision in the concreteness of the human body. Meanwhile, the "pencil of fire" inscribes a secular yet glorious universe henceforth ruled by the human being.

During this period, the study of optics undergoes significant changes. According to Crary, whereas vision had previously been understood as a privileged means of objective and disembodied knowing, it now turns into an object of empirical, experimental, and physiological knowledge. Goethe's *Farbenlehre* (1810) and Schopenhauer's *Über das Sehen und die Farben* (1815), Crary suggests, are crucial instances in the transition from the Newtonian, transcendentalist conception to a physiologically based one. Optical phenomena such as retinal afterimages, peripheral vision, binocular seeing, and thresholds of attention are subjected to empirical study, and a wide range of apparatuses and gadgets were developed to this end.

Significantly, these optical devices also become part of the popular culture. They are explored commercially in the form of widely popular toys, among them the thaumatrope, the phenakistoscope, the zoetrope, and the stereoscope. It is a double-faced process, Crary maintains. On the one hand, the observer is emancipated, and the autonomization of vision is affirmed in cultural practices such as optical diversions, photography, and painting. On the other hand, the scientific articulations of the observer's physiology are made compatible with new arrangements of power that emerge in the period, in particular the science of labor and the related rationalization of the human body.[25] Thus the scientific interest in the empirical observer functions as a system of power/knowledge, to use Michel Foucault's terminology.

The modernist period thus signifies the reinvention of the observer, now empirical, embodied, and particular, which helps explain why, as Martin Jay among others has maintained, vision loses its legitimacy as a philosophical metaphor and epistemological category. God has long since closed his infallible, transcendental eye; and the human eye, embedded as it is in the physical and anatomical functioning of the empirical body, turns out to be a poor substitute. In short, seeing is no longer necessarily knowing. In the realm of late nineteenth and early twentieth-century philosophy, according to Jay, the critical turn manifests itself in three ways: as a detranscendentalization of perspective, a recorporealization of the cognitive subject, and a revalorization of time over space.[26]

Mann's optical toys thus situate the world of the novel in a space of physiologically based vision, emancipated and instrumentalized all at the same time, a space, in short, haunted by the asymmetrical relationship between the object of perception and the perceiving subject. As a theme, this unbalanced dichotomy will resound in later sections of the narrative, prefiguring, in particular, Castorp's experience of how his body's opaque interior is articulated by the X-ray.

The Libidinal Look

The primary instance of emancipated vision is the eroticized gaze, weaving itself through the entire plot. What makes Castorp fall in love with Chauchat is above all her "narrow Kirghiz eyes," which, like painful yet pleasurable stigmas, conjure up his frustrated homoerotic love for his former schoolmate Pribislav Hippe and his "Slav" eyes.[27] For Castorp, every meal becomes an occasion for visual pleasure, and soon it is mutual:

> After Frau Chauchat had turned toward his table two or three times at meals [...] and each time found her eyes met by Hans Castorp's, she looked his way a fourth time on purpose—and met his eyes again. The fifth time, she did not catch him looking at her; he was not on guard at just that moment. And yet he immediately felt she was looking at him and turned his gaze so eagerly to her that she smiled and glanced away. [...] And so at the sixth opportunity, when he felt, sensed, knew somewhere deep within, that she was looking his way, he pretended to stare with emphatic distaste at a pimply lady who had stepped up to his table to chat with the great-aunt; he held his eyes fixed on her for a good two or three minutes, never yielding until he was certain that the Kirghiz eyes across the way had given up—a strange bit of playacting that Frau Chauchat could easily have seen through, indeed was meant to see through, so that Hans Castorp's refinement and self-control would give her pause. (*MM* 139–40/*Z* 196–97)

After so many rounds of obsessive dinner-table glancing, Chauchat deliberately drops her serviette, but Castorp's reaction displeases her. She punishes him by withdrawing her gaze. For two whole days Frau Chauchat never looks his way (*MM* 140/*Z* 197). This, if anything, confirms that Castorp has entered into an eroticized visual exchange with Chauchat. Later, it is time for Castorp's first medical examination. At lunch, he tries to catch a glimpse of Chauchat. Their nonverbal yet full-fledged intimacy has intensified to the degree that Castorp is now able actually to read her gaze:

> The charming patient had slowly turned her head, and a little of her upper body, too, to gaze plainly and openly over her shoulder at Hans Castorp's table—and not just at the table in general, no, quite unmistakably and very personally at *him,* a smile playing on her closed lips and in her narrow Pribislav eyes, as if to say: "Well? It's time. Are you going to go, Hans?" [. . .] The incident had confused and shocked Hans Castorp to the depths of his soul; he had barely been able to believe his eyes and had first gaped in stupefaction at Frau Chauchat's face, and then, raising his gaze above her brow and hair stared, had stared into space. (*MM* 173/*Z* 242–43)

When, finally, Castorp attempts a conversation with Chauchat, it happens, typically, on Walpurgisnacht, when the patients have staged a carnivalesque masquerade. For all his previous silence, Castorp now must speak. Indeed, he speaks furiously—in French, incidentally, just like Freud did when approaching delicate matters in his famous Dora case. Confessing his passion without further ado, he insists that he address Frau Chauchat with the informal "tu." Somewhat amused, she reveals that she will leave the following day, thereby terminating their specular relationship. Stunned, he prays: "Seven months beneath your gaze—and now, when I've come to know you in reality, you tell me you're leaving! [*Sept mois sous tes yeux . . . Et à présent, où en réalité j'ai fait ta connaissance, tu me parles de départ!*]" (*MM* 333/*Z* 465)

The masquerade evening, the narrator intimates, culminates in a sexual encounter between Chauchat and Castorp. Since she is about to leave, she furnishes him with a token that will replace her in her absence and enable him to continue their specular relationship while she is gone. His keepsake consists of a plate of glass, her X-ray plate (*MM* 343/*Z* 478). Mann made sure that the irony was not lost on his readers. During Chauchat's absence, Hans frequently presses his lips against the plate, as though it were an icon. Here, then, the amorous gaze is peculiarly refracted through the technological, creating yet another zone of ambivalence and impurity. Indeed, the X-ray plate is pulled out of its naturalizing habitat and estranged, no longer subject to *theoria* alone, but also to bodily *aisthesis*. This ambiguity resounds in other parts of the narrative, where specular activities are brought into play.

The Visual Hermeneutics of Class

Mann's novel, as I have suggested, stages the problem of *Bildung* and modes of subjectivity in terms of visually signifying systems. On closer consideration, it becomes apparent that the recurring, and seemingly metaphysical, question of the nature of the human being revolves largely around the new ways in which medicine and optical technology discover, map out, and so rearticulate the nature of the human body. Mann's narrative may be thought of as an experimental space, where at least three visually determined modes of knowledge intersect, confront, and compete with one another. The first system is social scanning, or the visual hermeneutics of class; the second is the medical gaze, or medical hermeneutics; the third, finally, is the mechanical eye of the X-ray machine in the chapter titled "My God, I see!"

In its classic, historically strong form, the identity of bourgeois man is contingent on the stability of a socially signifying system. Such a system is largely (but, of course, not only) visual, and all the more so in societies with strict class divisions; turn-of-the-century Germany is a case in point. The visual hermeneutics of class carries a set of institutionalized, yet semiconscious, social categories and maps it onto the exterior of the individual to be decoded. Mann's narrative provides numerous indications of how once stable socially signifying systems begin to disintegrate, how an antiquated notion of the bourgeois individual is replaced by a more functional, hence "modern," conception.

In order to grasp the implied parameters of the social gaze in *The Magic Mountain*, it is instructive to compare its hermeneutics with that of the photographer August Sander. In 1929, Sander declared that he was in the process of compiling a photographic portrait manual that he had begun before the war and that would provide the summa of the German people in the manner of an encyclopedia.[28] Sander's sociological worldview began with representatives of the agricultural world and "earthbound man," worked its way toward the industrial metropolis and its middle-class and upper-class inhabitants, and then proceeded to the other end of the curve: social outcasts, idiots, and beggars. Characteristically, Sander's aesthetics built on what could be designated as an *expressive physiognomy,* that is, the idea that the external representation of a person expressed the internal character, and that one's occupation—whether a farmer, an industrialist, or a ragpicker—emanated from one's personality (figure 2). But the assumptions underlying this expressive physiognomy revealed an already obsolete view of the organization of the social order. Indeed, Sander's encyclopedic endeavor was itself a symptom of social crisis.

The need for typologies such as Sander's typically arises in the wake of social changes, here those of industrialization. "It is no accident," Ulrich Keller remarks, "that Sander never took notice of the conveyor belt, that highly significant innovation in contemporary production methods, which most directly contradicted his antiquated ideal of work."[29] Only at the end of the nineteenth century did Germany make the transition from a predominantly rural state to an industrialized society, but this process then penetrated German society and culture all the more quickly. Ernst Bloch, in 1932, came up with the notion of *Ungleichzeitigkeit*—noncontemporaneity—in order to describe the peculiar social, temporal, and psychic consequences of the uneven character of this historical process.[30] Symbolically enough, Sander's project was never completed. The continuous social differentiation that triggered the project—it carries the emblematic title *Menschen des 20. Jahrhunderts*—exploded the attempt to construct an enclosed circle of social knowledge.

At the beginning of *The Magic Mountain,* the aesthetics of the expressive physiognomy is everywhere implied, albeit far from embraced. Consider, for instance, the portrait of Castorp's stout grandfather, the old-fashioned senator. Quiet, solid, dignified, his bourgeois outward appearance matches his station in life as well as his inward character. His life was built on work, an inherited calling, and a sense of duty. Just as his character manifests itself in his immediate environment—his house, furniture, belongings—so the exterior expresses his interior. Yet alienation has already made itself felt. Metaphorically speaking, the effects of the conveyor belt are implicit, but no less palpable; the ennui that haunts Castorp is precisely the spleen that fills that particular psychic space flung open when the historical links between occupation and identity have been severed, when the claims of the patrimony, and hence of the past, have forever been laid to rest.

The parameters of social scanning are established in a typical scene early in the novel. Castorp is still new to the establishment when he first encounters Settembrini, the Italian scholar and man of letters who will

2. August Sander, *Savings Bank Cashier, Cologne*, 1928. J. Paul Getty Museum, Los Angeles. © 2001 Die Photographische Sammlung / SK Stiftung Kultur—August Sander Archiv, Cologne / ARS, NY.

become a key figure in his intellectual adventures at the sanatorium.[31] Typically, Castorp takes the authority of his German social hermeneutics for granted, moving securely within its horizons of social decoding, at least for a little while yet. No episode is more telling than when he studies Settembrini's physiognomy. Focusing on what are, for him, immediately intelligible signs of the stranger's identity, the episode renders Castorp's social gaze and its wavering between many interpretive hypotheses:

It would have been difficult to guess his age, but it surely had to be somewhere between thirty and forty, because, although the general impression was youthful, he was already silvering at the temples and his hair was thinning noticeably, receding toward the part in two wide arcs, making the brow even higher. His outfit—loose trousers in a pastel yellow check and a wide-lapelled, double-breasted coat that was made of something like petersham and hung much too long—was far from laying any claim to elegance. The edges of his rounded high collar were rough from frequent laundering, his black tie was threadbare, and he apparently didn't even bother with cuffs—Hans Castorp could tell from the limp way the coat sleeves draped around his wrists. (*MM* 54/*Z* 79–80)

The description gives the reader an idea of the sophistication of Castorp's hermeneutics of class as he applies it to Settembrini's exterior: a sloppy coat, a worn-out collar, a shabby cravat, and no cuffs. In spite of these indexes of a certain poverty, however, Castorp detects signs that, in his view, imply that a gentleman stands before him after all: "the refined expression on the stranger's face, his easy, even handsome pose left no doubt of that" (*MM* 54/*Z* 80). Still, he is not entirely certain what to conclude from the sum total of these signifiers; indeed, his interpretive hypotheses refuse to enter the hermeneutic circle.

The question of nationality then comes to his aid. As a new cartography spreads out before him, it becomes clear to the reader that Castorp's hermeneutics of class easily translates itself onto a kind of ethnography:

This mixture of shabbiness and charm, plus the black eyes and a handlebar moustache, immediately reminded Hans Castorp of certain foreign musicians who would appear in his hometown at Christmastime and strike up a tune, then gaze up with velvet eyes and hold out their slouch hats to catch the coins you threw them from the window. "An organ-grinder!" he thought. And so he was not surprised by the name he now heard as Joachim got up from the bench somewhat flustered and introduced him, "Castorp, my cousin—Herr Settembrini." (*MM* 54–55/*Z* 80)

Thus it suddenly dawns on the German engineer what kind of social position the velvet-eyed Italian must occupy: he is an organ grinder! Nothing could be further from the truth, however: Settembrini is a distinguished private scholar, who celebrates reason, enlightenment, and Western progress. No doubt it is Settembrini's Italianness, combined with his loud, worn-out attire, that makes the young man jump to conclusions.[32] Castorp's hasty deduction follows the mental schemata of the opposition between *Kultur* and *Zivilisation,* a geospiritual map of Europe with Germany at the center, to which Mann, among many others in this period, makes constant allusions, both in his fiction and in his nonfiction. *Reflections of a Non-Political Man* (1918), which Mann wrote while working on *The Magic Mountain,* is a striking example. Briefly, this geospiritual map has two faces. Germany is seen either as the structural antidote to France—sometimes England also serves as the opposite—or as the perfect balance between East and West, North and South. As *das Volk der Mitte,* the German people is believed to mediate between barbarism (Slavic countries) and overrefinement (France/England), or between stout rationalism (the North) and soulful irrationality (the South, in particular Italy).[33]

Traditionally, "civilization" has referred to the self-consciousness of the western world in general, particularly with regard to its technological achievements. In German usage, by contrast, *Zivilisation* has been seen as secondary to *Kultur* and was perceived as such well into the twentieth century. As Norbert Elias has argued, the antithesis between *Kultur* and *Zivilisation* is linked to the German middle class and its rise to historical hegemony.[34] Beginning with Kant, *Kultur* is understood in contradistinction to *Zivilisation.*[35] Associated with nature, inwardness, depth, sincere emotions, immersion in books, and the cultivation of the individual, *Kultur* is placed in sharp contrast to *Zivilisation* and the courtly values of *politesse,* formal conversation, and ceremony. When, however, the German middle class becomes a vehicle of national consciousness, the opposition between *Kultur* and *Zivilisation* is translated into a cultural geography with expressly political overtones. The cluster of meanings surrounding the notion of *Zivilisation* then attaches itself to the image of France or England.[36] Moreover, *Zivilisation* increasingly denotes technological progress.

The second point to stress is the close connection between *Kultur* and *Bildung.* Goethe is typically regarded as one of the most significant champions of *Kultur,* active in an era that saw the ascendancy of the bourgeois class. Mann, on the other hand, is often seen as Goethe's intellectual heir in that period when the golden years of the bourgeoisie were drawing to a close. Georg Lukács was right, in other words, when he claims that the fundamental question haunting Mann's works, especially *The Magic Mountain,* is not so much "What is man?" as "What is bourgeois man?"[37]

If, then, we conceive of the bildungsroman as a formation of the specifically bourgeois subject, as a way of being that is "functional" in relation to a particular worldview, then a new reading becomes possible: *The Magic Mountain* is a story about how one mode of subjectivity replaces another. What is at stake in Mann's educational novel is the male middle-class body. The constitution of Castorp's identity, as we shall see, becomes a matter of the somatization of the self.[38] And the staging of this drama is, to use a Foucauldian phrase, predicated on the sanatorium as an essentially disciplinary space.[39] This, then, is yet another sense in which *The Magic Mountain* is to be understood as an experiment in identity formation. The sanatorium offers a space where a limited number of characters encounter and define one another in ways that are not socially possible in the world below, all destined to enter into a heterogeneous machinery of knowledge that explores and rearticulates their bodies and, by implication, their selves.

Let us return now to Castorp's reading of the physiognomy of the Italian intellectual. His hypothesis—that Settembrini is an organ grinder—reflects the social codes of the well-bred turn-of-the-century German bourgeois, which explains why Italianness signifies lightweight soulfulness and smiling music sold for a trifle. At the beginning of the novel, Castorp's point of view typically operates not as one perspective among others but as a transparent norm—at least as he sees it. This impression is overturned, however, as soon as Settembrini addresses him. Indeed, *The Magic Mountain* may be read as a narrative about how Castorp's worldview is relativized by other kinds of visually determined signifying systems that impinge on the conception of "man" as played out in the novel. Two pivotal moments may be distinguished, both bearing on Castorp's self-image and worldview: his encounter with the medical authorities at the sanatorium and the medical gaze in general, and his encounter with the X-ray machine and its mechanical eye.

The Epistemology of the Medical Gaze

When Castorp first meets with the two doctors who run the sanatorium, he attempts to decode the ulterior signs of their status. The sovereignty of his perspective, however, will soon be upset. For when the chief physician, Dr Behrens, looks at him in return, the doctor makes sure to assert the power of his gaze. This is still social scanning in the sense that the doctor engages in a reading of Castorp as a social being: "I spotted it at once. [. . .] There's something so civilian, so comfortable about you—no rattling sabers like our corporal here. You would be a better patient than he, I'd lay odds on that. I can tell right off whether someone will make a competent patient or not [*Das sehe ich jedem gleich an, ob er einen brauchbaren Patienten abgeben kann*]" (*MM* 44/*Z* 66).

Castorp, however, clings stubbornly to his identity as a visitor and engineer. He refuses to submit to the objectifying nature of the doctor's gaze—but in vain. Seconds later, the equilibrium is broken. Without further ado, the physician touches the engineer's body. He more than violates Castorp's physical autonomy; he challenges Castorp's right to be the center and author of his perception. The doctor does not touch a particular extremity, nor does he settle for the pulse; instead, he pulls down Castorp's eyelid and scrutinizes what he sees (*MM* 45/*Z* 67). In casting his glance on the young man's eye, the physician turns the most crucial instrument of perceptual knowledge into an object of knowledge. Vision, the noblest of the senses, makes possible classification and typologization, and so the doctor pronounces Castorp sick: "No doubt of it, totally anemic, just as I said. Do you know what? It was not all that stupid of you to leave your Hamburg to fend for itself for a while" (*MM* 45/*Z* 67).

This seemingly undramatic incident anticipates the later course of the novel. As Castorp's means of vision is turned back on itself, he is no longer just an engineer, someone who knows, but rather a phenomenon to be studied empirically, clinically, and theoretically. And as Mann's hero enters into an asymmetrical relationship to himself, where he, as a subject of knowledge, thinks about himself as an object of knowledge, as an organic being before the scrutinizing medical eye, he comes to inhabit an unbalanced dichotomy between object and subject. He begins to occupy a world of alterity, allegory, and difference. In short, he enters a world of interpretosis.

The next crucial step in the protagonist's self-formation is when he discovers that he has a fever. The narratological function of this discovery is, of course, to set the plot in motion, but it also leads to a series of events that give rise to interpretive confusion on the protagonist's part. Castorp has to take his temperature in order to "verify" his bodily state. It is his first experience using a thermometer. The visual translation of his interior is represented as a problem of legibility:

He was not immediately the wiser. The sheen of the mercury blended with the refraction of the light in the elliptical glass tube; the column seemed now to reach clear to the top, now not to be present at all. He held the instrument close to his eyes, turned it back and forth—and could make out nothing. Finally, after a lucky turn, the image became clear; he held it tightly and hastily applied his intellect to the task. And indeed Mercury had stretched himself, very robustly. The column had risen rather high, it stood several tenths above the limit of normal body temperature. Hans Castorp had a temperature of 99.7 degrees. (*MM* 166/*Z* 234)

Once the fever is an established fact, that is, once it has been translated into a visual and numeric record, Castorp informs Behrens about it.[40] The doctor responds, "And I suppose you think that's news to me, do you? Do you think I don't have eyes in my head?" (*MM* 171/*Z* 240). Castorp must now subject himself to a physical examination, revealing himself half naked before the medical gaze. Thus stripped of his bourgeois appearance, he offers himself to another visual horizon of interpretation, the medical one:

"Ah yes, it's *your* turn now! [*Ach so, das wären nun Sie!*]" he said; grabbing Hans Castorp's upper arm with one massive hand, he shoved him into place and gave him a sharp look. But he did not look directly at his face, the way you look at another human being, but at his body. He spun him around, the way you spin an object around, and examined his back as well. "Hmm," he said. "Well, let's have a look at what you're up to." And he took up his thumping again. (*MM* 176/*Z* 247).

During the examination, the doctor proceeds by way of percussion (tapping) and auscultation (listening), attempting to detect signs of pulmonary disease: dullness, rough breathing, scars on the lungs, pathological lesions, "moist spots," and so on.[41] Castorp's body spreads out as a vast legible surface before the physician's eyes.

The medical gaze, Foucault argues in *The Birth of the Clinic* (1963), represents a historical threshold in the history of medicine.[42] Before the emergence of modern theories of pathology, the physician rarely touched the patient, much less did the latter appear naked before the doctor—only after the patient's death would the physician have attempted to localize the site of disease. With the advent of pathological anatomy, however, the structured medical perception of the corpse is mapped onto the living body. The depths of the human interior turn into an extensive surface of legibility, to be deciphered by the medical gaze, a gaze that contains within itself a multisensory structure: sight, hearing, touch.[43] For Foucault, this perceptual and conceptual move signifies that the *forms* of visibility have changed. The new paradigm thus represents an "epistemological reorganization of disease in which the limits of the visible and the invisible follow a new pattern; the abyss beneath illness, which was the illness itself, has emerged into the light of language."[44] Foucault also maintains that clinical pathology had a tremendous impact on the constitution of the sciences of man, not just methodologically but also ontologically. Medical knowledge now had to balance the asymmetrical dichotomy of modern epistemology, which Foucault would later articulate in *The Order of Things* (1966): on the one hand, man is the object of knowledge; on the other hand, his consciousness is the transcendental condition for all knowledge.

The historical period Foucault covers in *The Birth of the Clinic* ends circa 1820. It is around this time that medicine becomes increasingly dependent on technological devices that transcribe invisible bodily facts into either auditory representations (the stethoscope) or visual representations, including graphic and numeric ones (photography, thermometry, sphygmography, radiography, and so on). To be sure, these devices may be theorized as extensions of the physician's already instrumentalized sensory organs, as prosthetic supplements. Loosely following Foucault's theory of the medical gaze, however, it seems more appropriate to conceive of these devices and their accompanying networks as parts of another shift in the history of medical perception. Once a device such as the X-ray machine transforms itself into a black box, in Bruno Latour's sense, it relegates the medical gaze to the realm of merely subjective, nonobjective, or not-yet-objective knowledge.[45] The machine thus helps to produce a new discursive realm of the subjective in which the once functional medical gaze is now localized. And as it splits open a new distinction between the subjective and the objective, the X-ray machine introduces new matrices of legibility and criteria of positivity, ultimately upsetting previous parameters of truth.

In tapping on Castorp's chest, Behrens thus applies the medical gaze to the body before him, a body whose depths, now legible, are projected onto its surface. As a consequence of the physician's not-yet-objective detection of a "moist spot," Castorp undergoes an X-ray examination, an event that again modifies his notion of self, now far more radically. Moreover, it introduces him to the enigmas of physiological perception. The opaque interior of the human body is suddenly brought to light by a language that names what is seen.

The Mechanical Eye

The International Sanatorium Berghof is a high-tech establishment, in particular because of its X-ray machine. The diagnostic device is represented both as an extension of the physician's senses and as an independent form of vision, uncannily endowed with its own agency. On the most obvious level, the X-ray examination confirms that Castorp is sick, and the illness gives him a new identity, in the most profound and wide-ranging sense of the word.

On another level, however, the X-ray machine is more than a single, if significant, narrative event. It enters the novel as a network, in Latour's sense.[46] The X-ray machine is an extended technogram—tubes, screens, lenses, switchboards, measuring instruments, cameras, plates. The point is that these artifacts have real consequences beyond the immediate laboratory, indirectly producing trajectories that link persons to persons, persons to things, or things to things. The X-ray plate, for instance, mediates between Castorp and Behrens, since it signifies the former's alleged disease and entitles him to stay at the Berghof. It mediates also between Castorp and Settembrini, since it provides a pretext for the latter's discourse on his skepticism with regard to the natural sciences, his disgust at Eastern inertia, and his involvement in the League for the Organization of Progress. As we have seen, the X-ray also connects Castorp to Chauchat, since she gives him a plate of her chest as a token to behold in her absence. Furthermore, the X-ray machine is a sociogram: it comes complete with doctors, nurses, technicians, and laboratory assistants. These extended multidimensional networks are essential for both plot and narrative.

When Mann conceived *The Magic Mountain,* X-ray technology was still relatively new. Röntgen, a German physicist, had discovered the X-ray in 1895. Within days, the news of rays that were "entirely invisible to the eye" was cabled all over the world.[47] While experimenting with a cathode-ray tube and electrical discharges through rarefied gases, Röntgen had observed a mysterious fluorescence. He noticed that the rays penetrated solid matter. By placing books and various metals between the tube and a screen of barium platinocyanide, he concluded that the denser the metal, the more radiation was absorbed. Encyclopedias of the history of science often relate what henceforth would seem to be the moment of discovery: when Röntgen put his hand between the tube and the screen, the bones in his hand appeared on the screen.[48]

Shortly after his discovery, Röntgen presented a paper on this unknown type of electromagnetic radiation of short wavelength. He sent the paper to influential physicists, along with X-ray prints, including one of his wife's hand. Soon the photograph of Frau Röntgen's hand bones, complete with an opaque wedding band, made its way into scholarly journals, newspapers, and the popular press, in Europe and in the United States. Its

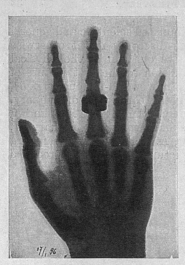

FIG. 12a —Roentgen picture of a hand, made on January 17, 1896, by G. Klingenberg and Slaby in Charlottenburg.

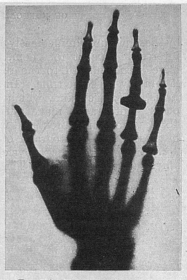

FIG. 12b.—Roentgen photograph of a woman's hand made in January, 1896, by P. Spies in Berlin.

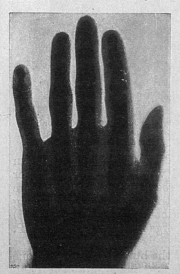

FIG. 12c.—Shadowgraph of a living hand, made in January, 1896, by A. A. C. Swinton and J. C. M. Stanton in London.

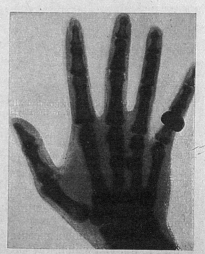

FIG. 12d.—Radiograph of a hand with bullet, made in January, 1896, by A. Londe in Paris.

3. The X-rayed hand as a new photographic genre. Plates from Otto Glasser, *Wilhelm Conrad Röntgen and the Early History of the Roentgen Rays* (Springfield: C. C. Thomas, 1934), 34–35.

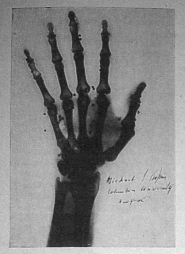

Fig. 12e.—Hand with shot. Photograph made in February, 1896, by M. Pupin in New York.

Fig. 12f.—Human hand. Picture made in February, 1896, by G. S. Moler in Ithaca, N.Y.

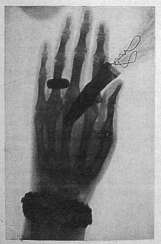

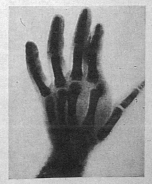

Fig. 12g.—X-ray photograph of hand with bouquet of flowers, made in February, 1896, by W. König in Frankfurt-a.-M.

Fig. 12h.—Roentgen picture of boy's hand showing trauma of bone, made January 29, 1896, by W. König in Frankfurt-a.-M.

caption read *Hand mit Ringen*. As Otto Glasser has observed, the x-rayed hand became a photographic genre unto itself, appearing in both scientific journals and popular publications (see figure 3).[49]

Even more important, the X-ray helped reconfigure the conception of the human body.[50] As Stanley Reiser suggests in *Medicine and the Reign of Technology*, the X-ray obliterated one "distinction between the outer and inner spaces of the body—both were now susceptible to visual examination." The general interest revolved around the recognition that the X-rays revealed the skeleton, making the surrounding flesh look like a halo; similarly, the X-ray appeared as a "materialized eye."[51] Just how widespread the public interest was may be gleaned from the fact that X-ray machines and prints were displayed in shop windows. Equally astonishing, X-ray photographs became tokens of sentimental value; New York women of fashion had their hands photographed in order to prove that bone structure makes a hand beautiful; and an X-ray print of mother's hand became a popular family souvenir.[52]

Clearly, then, Castorp was not alone in fetishizing X-ray plates. Cherishing the print of Chauchat's chest plate, he keeps it in his breast pocket next to his heart and often kisses it. In all likelihood, Mann's contemporary readers must have recognized Castorp's charged relationship to "the funereal photograph [*funebre Lichtbild*]," and Mann evidently alludes to a readily available popular discourse surrounding X-ray technology (*MM* 238/*Z* 332). Mann's term—*funebre*—confirms that this discourse was suffused with a rhetoric of death; radiographic images, we recall, were often referred to as "ghost pictures" or "shadowgraphs."

From the point of view of medicine as a discipline, the emergence and use of radiography had not only methodological bearings but also theoretical ones. Understood as an amalgamation of technical forces and specialized human skills, the X-ray machine may be seen as part of a technogrammatic network that helped restructure the matrices of medical perception and the implicit rules that govern the diagnostic gaze. Not only did the physiological interior become subject to new forms of legibility; the X-ray also helped reorganize the epistemological assumptions on which the medical gaze previously rested.

What is more, because the X-ray translated the bodily interior into a visual record, it provided a seemingly true-to-life representation that was permanent and that could be stored. If the symptoms and other signs of disease previously were inseparable from the individual body of the patient, it was now possible to mobilize them as autonomous items in their own right, in the form of objectified visual inscriptions. This also allowed the patient to carry around the exteriorized visual sign of his or her own interior.

The Legibility of the Interior

Mann's novel renders an individual's first encounter with X-ray technology. In this respect, too, Castorp is ever the neophyte—no one has ever "taken a look [*Einblick*] into his organic interior [*Innenleben*]" (*MM* 207/*Z* 290). "I believe," Dr Behrens says, "you're afraid to reveal your insides to us, aren't you, Castorp?" (*MM* 212/*Z* 296). While Castorp sits in the waiting-room adjacent to Behrens's laboratory, Chauchat slips in unexpectedly. Castorp, who has yet to introduce himself to her, has already had erotic fantasies about her. Now that the object of his daydreams materializes in person, the reader realizes, Castorp feels somewhat embarrassed. Castorp suspects that Chauchat is having an affair with Behrens, who has painted her portrait in the privacy of his studio. Scanning her body, posture, and clothes, Castorp's gaze finally comes to rest on her chest, a body part that triggers some contemplative activity: "Suddenly Hans Castorp recalled that she was also here waiting to be X-rayed. The director was painting her, interpreting her external appearance with color and oils on canvas. But there in the twilight, he would turn rays on her that would expose the inside of her body. And at the thought, Hans Castorp turned his head to one side, and his face darkened with the shadow of respectability and assumed a look of discretion and propriety that seemed appropriate to such a vision" (*MM* 210/*Z* 295).

In a striking structural parallel, Behrens, the physician-turned-painter, is now to render her organic interior. The passage suggests that as Castorp reflects on this fact, the boundary between interior and exterior, surface and depth, begins to blur. If it is rather indecent to fantasize about Chauchat naked, then what does it mean to picture her being X-rayed, that ultimate, yet noninvasive, bodily penetration? Why would it be more improper to visualize her surface, that is, her nakedness, than her depth, that is, her skeleton and organs? If her flesh is really a mere surface, enveloping a visible invisibility, are not the standards on which bourgeois decorum rests somewhat arbitrary?

In the X-ray laboratory, Castorp looks on while his cousin is examined, and the metaphorical eye is evoked once again. The event is rendered as a spectacle, whose terms are framed by the notion of the technological sublime:[53] "For two seconds the dreadful forces necessary to penetrate matter were let loose [. . .]. Barely tamed for their purpose, these forces sought other outlets for their energy. Discharges exploded like gunshots. The gauges sizzled with blue light. Long sparks crackled along the wall. Somewhere a red light blinked, like a silent, threatening eye, and a vial behind Joachim's back was filled with a green glow" (*MM* 212/*Z* 297–98).

When Castorp undergoes examination, the sublime is folded back into its traditional realm—nature and landscape—for the X-ray enlightenment is likened to a storm and is thus apprehended in terms of nature. The noninvasive penetration of the interior, by contrast, is represented as a scandal of the mind, a contradiction of the first rank: "Hans Castorp waited, his eyes blinking, his lungs full of air. The thunderstorm burst behind him, hissing, crackling, popping—and fell quiet again. The lens had peered inside him [*Das Objektiv hatte in sein Inneres geblickt*]. He dismounted, confused and dazed by what had happened to him, although he had not felt anything at all during the penetration" (*MM* 213/*Z* 298).

After Behrens has taken X-rays of the two young men, he moves to a different apparatus, which, although not specified in the text, is probably a fluoroscope, since the interior organs can be seen on a screen. Alluding once again to the enigmas of sight, Behrens intimates that the object of visual perception is contingent on the physiological functioning of the eye itself. Implicitly, the question emerges as to whether the object seen exists independently of the perceiving subject: "We have to wait for our pupils to get nice and big, like a cat's, in order for us to see what we want to see. I'm sure you can understand that we can't see properly, just like that, with our normal daylight eyes. For our purposes here, we first have to ban any rousing daylight scenes from our minds" (*MM* 213/*Z* 299).

A motor starts, the floor vibrates, and the red light penetrates Joachim's body. Worried that he might commit an indiscretion, Castorp asks his cousin's permission to look at his insides. Flabbergasted, he gazes into the opaque interior illuminated with the aid of "physical optics [*physikalisch-optischen Wissenschaft*]" (*MM* 215/*Z* 301). Behrens points at the screen: "Sharp picture," he exclaims. "Do you see the diaphragm? [. . .] Do you see the hilum there? Do you see those adhesions? Do you see these cavities here?" (*MM* 214/*Z* 300). But Castorp is incapable of reading what he sees. What is more, his attention is drawn to a baglike phenomenon, whose regular contractions make him think of a jellyfish. He is unable to distinguish any features, much less identify them. When, however, the doctor directs his means of vision and gives a name to the pulsating shadow, Castorp is made to understand that he beholds the symbol of life itself, the core of the human being: the heart. All of a sudden, his cousin's "dry bones, his bare scaffolding, his gaunt *memento mori*" is revealed before his baffled eyes: "Yes, yes! I see it [*Jawohl, jawohl, ich sehe*]," Castorp exclaims. "My God, I see it! [*Mein Gott, ich sehe!*]" (*MM* 215/*Z* 301).

This pivotal event epitomizes the ways in which the limits of the visible and the invisible, depth and surface, are rearranged. By the same token, the X-ray examination demonstrates how the limits of what one can know and not know are redefined. Proposing that seeing is reading, the episode indicates that vision depends on a language that names what is seen. A new epistemic space is thus opened up, a space produced by an epistemological reorganization of the interior of the human body.

But the X-ray session has a significant coda. Castorp asks to see his own hand pierced by radiography.[54] This incident produces an intertextual relationship between Castorp's ensuing revelation and the widely circulated story about Professor Röntgen's moment of discovery in 1895. We recall that Röntgen inserted his hand between the cathode-ray tube and the screen. We recall also that the print of Frau Röntgen's hand functioned as "evidence," as a repetition of this moment. The rendering of Castorp's visual experience is thus a symbolic reenactment of Röntgen's discovery, except that it is played out in an existential register:

Hans Castorp saw exactly what he should have expected to see, but which no man was ever intended to see and which he himself had never presumed he would be able to see: he saw his own grave. Under that light, he saw the process of corruption anticipated, saw the flesh in which he moved decomposed, expunged, dissolved into airy nothingness—and inside was the delicately turned skeleton of his right hand and around the last joint of the ring finger, dangling black and loose, the signet ring his grandfather had bequeathed him [. . .]. With the eyes of his Tienappel forebear [. . .] he beheld a familiar part of his body, and for the first time in his life he understood that he would die. (*MM* 215–16/*Z* 302)

The repercussions of this insight are more than existential, however, for the shock Castorp suffers is also ontological. The experience relates to his own being, specifically his being as an object of knowledge. Significantly enough, it is only after this traumatic insight that Castorp becomes what may be identified as an intellectual. Enabled by the mechanical eye of the X-ray machine, his altered relationship to his own being, now temporalized in the face of death, produces a radical will to know—which at the same time discredits all that he used to know in the world down below the sanatorium.[55] Castorp's will to know is produced by two events, and both are responsible for his ever longer stay at the sanatorium: first, his love for Chauchat, an unfulfilled desire that, in psychoanalytic terms, is a displacement of a previous, homoerotic, and equally unfulfilled relationship; and, second, the X-ray examination, which proves both that he is sick and that he will die. So it is that Castorp becomes an intellectual under the sign of lack.

Epistemophilia

According to Dr Behrens's reading, the X-ray plate indicates a "few nodular lesions" on Castorp's lungs. Behrens concludes that the young engineer is ill; and this, in turn, enables Castorp to establish his freedom. In the subsequent chapter, entitled "Freedom," he writes a letter to his relatives, explaining that he must remain at the Berghof. As he signs his name, he underwrites his existential freedom. Now a new life begins, betokened, among other things, by a new possession—the X-ray plate. As Settembrini confirms in his characteristically ironic manner, the X-ray plate has provided the young man with an identity: "Ah, you carry it in your wallet. As a kind of identification, like a passport or membership card. Very good. Let me see" (*MM* 238/*Z* 332).

In addition to the corporeal discipline at the sanatorium, Castorp's new life is one of bookish learning, intellectual discussion, and frustrated love—in short, the cultivation of the interior. Most of all, he finds himself overtaken by a tremendous urge to accumulate knowledge. Haunted by the question "What is life?" he begins an inquiry into the natural sciences—in particular, those dealing with the human body. Shortly after the X-ray episode, Castorp and his cousin pay a visit to Behrens's studio. The underlying motive is to take a good look at the portrait of Chauchat, the amateurish representation of which sparks a discussion of aesthetics—for example, how to render female plasticity in an anatomically correct fashion. Focused on the portrait of Chauchat, the discussion veers toward physiological theories of the skin, of the blood and the lymph, and Castorp becomes excited: "The flesh! The human body! What is it? What is it made of? Tell us now, this very afternoon, Director Behrens. Tell us, for once and for all, in precise terms, so that we may know" (*MM* 261/*Z* 365). But Behrens, overcome by an onslaught of melancholy, presumably triggered by the thought of death, brings the spirited exchange to a halt.

Subsequently, Castorp sets his mind on private study and orders a wide range of scientific works. His inquiry into the nature of life assumes the form of a search for foundations, whether in physiological and chemical theories of the cell, physical theories of the atom and the molecule, embryological theories of the ova and spermatozoa, or astronomical theories of the universe. As he lies on the balcony taking his rest cure, "the image of life" suddenly presents itself to him, almost like a celestial vision, and a mysterious creature appears before his eyes:

> The image hovered out there in space, remote and yet as near as his senses—it was a body: dull, whitish flesh, steaming, redolent, sticky; its skin blemished with natural defects, blotches, pimples, discolorations, cracks, and hard, scaly spots, and covered with the delicate currents and whorls of rudimentary, downy lanugo. The body was leaning back, wrapped in the aura of its own vapors, detached from the coldness of the inanimate world, its head crowned with a cool, keratinous, pigmented substance that was a product of its own skin. (*MM* 272/*Z* 380).

The obscure creature is the result of Castorp's delirious fantasy, projected onto the eroticized image of Chauchat. In subjecting her hair and other bodily features to physiological, anatomical, and neurological theorization, he produces a series of ecstatic reflections leading to the next topic on the intellectual agenda: procreation itself. Castorp reflects on sex cells, theories of ovulation, and embryology, thus inflecting his inquiry into the origins of organic life along pseudo-sexual lines. This episode emerges as a parallel to Castorp's visualization of Chauchat's X-rayed chest; here, too, medical imagination manages to sidestep bourgeois decorum.

It is interesting to note that for Freud, the *Wisstrieb*, or the epistemophilic drive, is intertwined with the *Schaulust*, or scopophilia, the eroticized desire to see.[56] According to Freud, the will to knowledge is elicited by the question, "Where do babies come from?" A significant part of Castorp's will to know may arise out of a similarly primeval question. In this way, the thematics of vision, visuality, and specular relationships that is so pronounced in Mann's novel takes on new and unexpected significance. *The Magic Mountain* lends itself to a psychoanalytic reading, where the decisive role of vision in epistemological scenarios might be explored in some detail—for, as we have seen, it is only after his confrontation with the issue of how vision relates to knowledge that Castorp acquires a new sense of self, including his bodily self.

Having completed his initial research into the enigmas of life, Castorp is pleased to conclude that he has traversed three discursive fields. Now he stands in a threefold relation to organic nature—"lyric, medical, and technical. It came as a great inspiration. And these three relationships, he believed, were a unity within the human mind, were schools of humanist thought, variations of one and the same pressing concern" (*MM* 276/*Z* 386). At the same time, Castorp has begun his project of self-formation, of *Bildung,* transforming his bodily being in the process of discovering it, the result being a transfiguration of the male middle-class body.

Rising to the Aesthetic

Yet Castorp's long process of *Bildung* is not complete until he has been introduced to a new technological device, the gramophone. The sanatorium has just acquired the latest model, endowed with an electric motor, and this "magic box" provides a new form of evening diversion. The narrator stresses that it replaces those old-fashioned optical toys discussed earlier: "there was no comparison to those little mechanisms in value, status, and rank" (*MM* 627/*Z* 872). But there is another form of outclassed visual pleasure, another downgraded regime of visual consumption, that emerges as a counterpoint to the gramophone and that, furthermore, reveals the contradictions inherent in Mann's account of Castorp's sublime musical experience: film. From a structural point of view, the advent of the gramophone is, as we shall see, both a supersession and an inversion of another aesthetic experience—the previous visit to the movie theater in the village. Film and recorded music: both are technologically mediated and reproduced. Both cater to an ever-expanding audience. Both reify the here and now of the "original" performance behind the "copy." Both are subject to endless repeatability. In *The Magic Mountain,* however, film is rendered as a downgraded art form—the question is whether it deserves the designation *art* in the first place—whereas gramophone music is subject to celebration and high esteem, albeit somewhat ironically.[57]

One day, Castorp and his cousin decide to take a mortally ill young woman to the Bioscope Theater. In the company of an anonymous mass of spectators, the trio watches an Orientalist film. On the screen, "life flickered before their smarting eyes [*schmerzenden Augen*]—all sorts of life, chopped up in hurried, diverting scraps that leapt into fidgety action, lingered, and twitched out of sight in alarm, to the accompaniment of trivial music, which offered present rhythms to match vanishing phantoms from the past and which despite limited means ran the gamut of solemnity, pomposity, passion, savagery, and cooing sensuality" (*MM* 310–11/*Z* 434).

The terms of this narrative event recall the charged rhetoric in which numerous historical accounts of first visits to the movie theater are couched: the flickering images, the ghostlike darkness, the uncanny nature of the shadowy figures projected onto the screen. The episode thus carries a paradigmatic weight, which is then affirmed when the narrator lets the event unfold into a miniature essay. Described in detail, the Orientalist film is rendered as cheap, vulgar, and crude, as it revels in a Moroccan woman's swelling cleavage and an executioner's muscular arms.[58] And because it seeks to gratify an international audience, the narrator emphasizes, the film suffers from a lack of artistic ambition and intention, a tendency intrinsic, it seems, to the medium itself. As if this were not enough, cinema is also characterized by its depersonalized mode of production and distribution: "There was no one there to clap for, to thank, no artistic achievement to reward with a curtain call" (*MM* 311/*Z* 435). This absence, the narrator intimates, is typically matched by the "repulsive" behavior of the spectators, a mass that is as passive as it is lacking in distinction and aesthetic judgment: "Once the illusion was over, there was something repulsive about the crowd's nerveless silence. Hands lay impotent before the void. People rubbed their eyes, stared straight ahead, felt embarrassed by the brightness and demanded the return of the dark, so that they could again watch things, whose time had passed, come to pass again, tricked out with music and transplanted into new time" (*MM* 311/*Z* 435).

Film, then, is not only a pure diversion but also a cultural practice whose production and reception are fundamentally mediated by the masses. Interestingly enough, Castorp's consumption of prerecorded music does not elicit the same reaction, although in that context, too, there is no one to applaud, no one to thank for the beautiful rendition. While demonstrating the gramophone to his baffled patients, Dr Behrens introduces a distinction between a musical instrument and a mere apparatus, as if to even up the machinelike edges of the new device. In the same breath, he alludes to the charged discourse of *Kultur* versus *Zivilisation*, asserting a close link between the essentially German and the technological: "This is no apparatus, no machine, [. . .] this is an instrument, this is a Stradivarius, a Guarneri—you'll hear resonances and vibrations of vintage *raffinemang*. It's a Polyhymnia, as we are informed here inside the lid. German-made, you see—we make far and away the best. Music most faithful, in its modern, mechanical form. The German soul, up-to-date" (*MM* 627–28/*Z* 873).

Perplexed, the patients listen to an Offenbach overture that, disembodied from its original performance, its particular presence in space and time, now offers itself abstractly, as it were, to their auditory organs. The narrator reports how the listeners perceive a change with regard to the sound volume, the *Klangkörper,* and how it suffers from a diminution of perspective: "it was, if one may use a visual comparison for an audible phenomenon, as if one were gazing at a painting through the wrong end of opera glasses, so that it looked distant and small, but without forfeiting any definition of line or brilliance of color" (*MM* 628/*Z* 874). Such an analogy testifies to the persistence of vision as metaphor or, one could say, to its reification as metaphor. But it also bears witness to the perceived freshness of technologically stored and transmitted musical sounds, an aesthetic experience that, at this point, awaits its particular sensory modes of management and processing.[59]

Castorp immediately acquaints himself with the gramophone, eventually becoming its prime operator. With all of the distinction bestowed on cultivated men, he listens to recorded music late at night all by himself. Unlike his visit to the movie theater in the village, Castorp's encounter with gramophone music is not a collective, mass-mediated experience. Indeed, his solitude impregnates these late-night aesthetic experiences, lending them a certain sublimity. Ultimately, then, it is the mode of distribution that sets film apart from recorded music and so degrades it.

It seems proper that the sluggish engineer-cum-intellectual should evolve into a distinguished and passionate listener, even if he has yet to advance to genuine Wagnerianism.[60] For Mann, as for Adorno, Nietzsche, and Schopenhauer, music represented the highest and most refined art form. Nietzsche, for one, maintained that a life without music would be equivalent to a life in exile. In his view, music was a basis for philosophy; the more one immersed oneself in the spiritual realm of music, the more one learned to think authentically.[61]

In this respect, then, Castorp's *Bildung* is more or less complete. He has successfully acquired the sensibilities associated with *Kultur:* bookish learning, emotional depth, aesthetic judgment, and inner enrichment. In short, he has acquired individuality. His ability to immerse himself emotionally and existentially in a Schubert *Lied* and its worlds of "feeling" confirms his spiritual graduation. Ernst Robert Curtius, in a 1925 review article of *The Magic Mountain,* goes so far as to read Castorp's musical experience as a feature of what is typically German. In fact, Curtius contends that Mann's work is deeply rooted in a mysterious realm—a Germanness whose essence is musical and metaphysical all at once (*musikalisch-metaphysischen Deutschtum*).[62] This notion, obviously, is what Behrens alludes to when he suggests that the gramophone offers an updated version of the German soul. The irony inherent in the physician's remark should not mislead us. In this way, Mann attempts to keep intact the notion of the German soul.

Apart from such messages, the most important lesson that the new technological device, variously referred to as "chest," "cabinet," "little truncated coffin of fiddlewood," "small dull-black temple," has in store for Castorp is memento mori. "What was this world that stood behind [the Schubert *Lied*], which his intuitive scruples told him was a world of forbidden love? It was death" (*MM* 642/*Z* 893). To the notion of *Kultur* is now added "sympathy with death," the ultimate form of "triumph over self." The chapter revolves around musical harmony, but it scarcely resembles the conclusive harmony of the classic bildungsroman. Castorp, after all, decides to go to war. Once again, then, the image of technology attaches itself to the image of death, as though the emergent technologies for reproducing art, including such ephemeral artforms as music, can only shatter the bourgeois world of *Bildung*, its inherent values and intrinsic modes of aesthetic perception. Mann's novel, as Jochen Hörisch has proposed, is the "swan song" of the educated reader.[63]

If, as I have shown, the experience of the machine is vital to the formation of the hero and to Mann's novel as a whole, why is it that these thematic complexes have gone largely unnoticed?[64] Surely, its role has gone unnoticed not because the machinery in the novel enjoys a peripheral status. On the contrary, and this may seem paradoxical, the significance of technology has been disregarded precisely because it is so central to the narrative. Against this background, it is instructive to compare Mann's novel with the surrealist preoccupation with technological artifacts. The surrealist machine is more often than not a nonfunctional machine, stripped of its living environment and naturalizing context. Concerning the surrealist predilection for fetishizing the machine, Hal Foster suggests that nearly "all machinist modernisms fix fetishistically on the machine as object or image; rarely do they position it in the social process."[65]

In *The Magic Mountain*, by contrast, the machine is placed in its natural habitat, and this is part of the reason why it has been overlooked. The X-ray machine enters the novel as a vast network, weaving its way through the entire plot, up until the point when Castorp receives his last X-ray plate toward the end of the narrative. The X-ray machine emerges as both a technogram and a sociogram; hence it is a working machine, that is, a naturalized technology. The X-ray machine is effective precisely because it has been incorporated into other practices, social, intellectual, economic, even emotional; it mediates these practices while it is, at the same time, mediated by them. In the final analysis, the space of the sanatorium naturalizes not only the machine but also the intellectual inquiries engaging Mann's protagonist. This, then, is the sense in which the sanatorium is not just a background, much less a mere symbol of an ostensibly moribund European culture. Indeed, the sanatorium, like the machine, is an integral part of the inner logic of the plot: its condition of possibility, its pretext, its edifice.

Mann and the Total Work of Art

We have seen the extent to which *The Magic Mountain* is preoccupied with various forms of specular pleasure, vision machines, and mechanical extensions of the human eye. We have also seen how these technologies are represented as reducing the habits of human vision to secondary, merely "subjective," experiential modalities. In yet another respect, then, it is fitting that the novel end with Castorp's elevation to the aesthetic. As he manages to ascend to the highest art form—music—he rises above the reified forms of vision mapped out in the novel.

Yet a certain historical irony casts its shadow over this final instance of Castorp's spiritual expansion, since the abstraction of vision also participates in the separation of the senses, ultimately encouraging their aestheticization and reification. Nietzsche, writing in 1878, argues that the production and reception of music are at that moment undergoing something like a rationalization process. In *Human, All Too Human* (1878), he notes apropos of what he called the desensualization of higher art, "By virtue of the extraordinary exercise the intellect has undergone through the artistic evolution of modern music, our ears have grown more and more intellectual. We can now endure a much greater volume, much more 'noise,' than our forefathers could because we are much more practised in listening for the *reason in it* [*Vernunft*] than they are."[66]

Following in Nietzsche's footsteps, Adorno subjects Wagner's musical dramas, his *Gesamtkunstwerke*, or total works of art, to critique. Wagner's explicit intention is to emancipate the ear. He also wants to transcend the genre system: the *Gesamtkunstwerk* is, Wagner hopes, the drama of the future, a total art form in which poetry, music, and theater would be united and transcended in a historically unprecedented way. In promoting the idea of the *Gesamtkunstwerk* as a synthetic form, indeed a synaesthetic art form, Wagner also expresses a will to reconcile the senses and dissolve their institutionalized aesthetic division of labor. For Adorno, however, such desire rings of deceptive wish-fulfillment. Hearing is the last not-yet-reified sense, not yet appropriated by the means-end rationality of bourgeois society. Yet to pursue the emancipation of the ear, as Wagner does, is to effect its rationalization, for in Wagner's musical dramas, hearing is singled out as that sense whose task it is to codify and process the linear development of the diegesis of the drama. In this way, Wagner's grand protest partakes in the process of reification precisely by means of trying to overcome it. Adorno warns that it is questionable whether there ever existed a unity of sense experience, the myth on which Wagner's musical dramas feed. The composer is therefore forced to dream transcendence at the price of inevitable failure. Synaesthesia is thus precluded: "The senses, which all have a different history, end up poles apart from each other, as a consequence of the growing reification of reality as well as of the division of labour. For this not only separates men from each other but also divides each man with himself."[67]

A similar logic is at work in *The Magic Mountain*. Not only does Mann elevate the musical leitmotiv into a structuring principle in the novel; he does so in an explicitly Wagnerian spirit. On the most obvious level, the leitmotiv technique is a mode of organizing the narrative materials. In fact, Mann once noted that the "conception of epic prose-composition as a weaving of themes, as a musical complex of associations, I [. . .] largely employed in *The Magic Mountain*. Only that there the verbal leitmotiv is no longer, as in *Buddenbrooks*, employed in the representation of form alone, but has taken on a less mechanical, more musical character, and endeavours to mirror the emotion and the idea."[68] Mann here produces a distinction between the mechanical and the artistic that, as we have seen, echoes widely in modernist art and criticism. More important, since the leitmotiv is a mode of organizing narrative materials and arranging a complex of associations, it is also a means of structuring readerly perception and reception. The leitmotiv thus turns into a modality for molding and constructing knowledge. So it is that *The Magic Mountain*, as befits an encyclopedic novel, transforms itself into a vast apparatus for organizing and producing knowledge.

Moreover, *The Magic Mountain* attempts to enclose within itself, even to exhaust, other media of cultural production with which it competes, in particular mass-cultural ones, such as optical diversions, cinema, color photography, and gramophone music. Each of these media of cultural production is subjected to detailed reflection, sometimes even to miniature essays, particularly film and gramophone music. Symptomatically, then, Mann's literary encyclopedia contains entries on almost all art forms, and, as we have seen, these entries come complete with an implicit axiology. In an effort to synthesize the various cultural modes of production and so outdo them all, *The Magic Mountain* thus turns itself into a giant aesthetic edifice: a literary counterpart of that total work of art, the *Gesamtkunstwerk*.

Mann's novel also encloses within itself competing modes of knowing. In the narrative, X-ray technology emerges as a mode of knowledge, or, more precisely, as a mode of inscription and translation that makes other modes of knowledge seem antiquated and in some ways arbitrary, especially the bourgeois socially signifying system and the medical gaze. At the same time, the novel attempts to present a solution to the contradictions arising out of these conflicting visual economies and signifying systems, and, crucially, projects it in the narrative form itself. *The Magic Mountain* effectively overarches the new and alien languages of medicine and technology, granting these idiolects but local validity within the larger universe of *Bildung* and the total work of art. (In fact, as Heinz Saueressig has observed, Mann was proud of the fact that even though the novel contained vast portions of medical discourse, specialists had found nothing to object to.)[69] As a total work of art, *The Magic Mountain* dreams of an ultimate sensibility: a humanism capable of subsuming the "lyric, medical, and technical," that is, a sensibility that knows no boundaries between art, technoscience, and experience. *The Magic Mountain* may then be seen as an attempt at overcoming precisely the contest of the faculties it so forcefully represents. Yet, at the very moment the novel aims at transcending the ever-widening gap between humanism on the one hand and science and technology on the other, it can only reinscribe their respective boundaries.

On another, more deep-seated level—that of the imaginary[70]—the narrative already admits that the grand moment of *Kultur* has passed. The recurrent images of death and human finitude that so strikingly coagulate around the cluster of technological apparatuses suggest a language set in another key, a language of fear, alarm, and nostalgia that ultimately runs counter to the desire for a synthetic, all-embracing humanism in a world of social, scientific, and technological interpretosis.

3 The Education of the Senses

Remembrance of Things Past and the

Modernist Rhetoric of Motion

Cinema is the materialization of the worst of popular ideals . . . At stake is not the end of the world, but the end of a civilization.
ANATOLE FRANCE

A train whistle cuts through the opening pages of Marcel Proust's *A la recherche du temps perdu* (1913–27), translated as *Remembrance of Things Past.*[1] Rushing through the night, the train rouses Proust's narrator from that nocturnal space-time he usually inhabits upon falling asleep. The train, however, triggers new fantasies. Recalling a distant world far beyond the narrator's bedroom, the whistle evokes parallel lives and alternate spaces as intriguing as those he entered upon dozing off. The very first sense impression recorded in Proust's novel is thus auditory and issues from the iron horse, that great mid-nineteenth-century novelty.

The train not only helped change habitual notions of time and space; it was also believed to produce previously unknown ailments, such as "railway spine."[2] Medical experts similarly worried about the passenger's perceptual organs; they cautioned that the intermittent views through the windows fatigued the passenger's eyeballs, just as the squealing brakes were believed to strain the ears. In Proust's novel, however, the train is a natural part of the landscape. Nothing could be farther from, say, John Ruskin's train, that industrial conveyor of human bodies whose velocity destroyed preindustrial modes of perceiving landscape; or Emile Zola's monstrous vehicle of violence and death in *La bête humaine* (1890); or, for that matter, Claude Monet's heroic puffers, steaming into La Gare Saint-Lazare in the eponymous paintings (1876–77).[3] In *Remembrance of Things Past,* it will take more futuristic means of transport—an airplane or an automobile—to create sensations in Proust's narrator that resemble the excitement, malaise, or agony that the railway had elicited in an earlier generation.

For the narrator, as he lies in his bed, the distant train whistle becomes instead the occasion for a pleasurable fantasy about exploring unknown territories, encountering strangers, and exchanging stories:

> I would ask myself what time it could be; I could hear the whistling of trains, which, now nearer and now farther off, punctuating the distance like the note of a bird in a forest, showed me in perspective the deserted countryside through which a traveller is hurrying towards the nearby station; and the path he is taking will be engraved in his memory by the excitement induced by strange surroundings, by unaccustomed activities, by the conversation he has had and the farewells exchanged beneath an unfamiliar lamp, still echoing in his ears amid the silence of the night, by the imminent joy of going home. (*REM* 1:3/*RTP* 1:3–4)[4]

The narrator has just rendered his bedtime habits, how he would go to bed, read for a while and doze off, only to wake up about half an hour later, looking to blow out the light that he has already put out and to close the book that he has already laid aside. Here, as always in Proust, genuine temporality is a property of the individual psyche. This means that the individual psyche, rather than any external chronology, is construed as the privileged site of temporality. But how can this be? Proust solves the problem by focusing on the body, turning it into a prime mediator between past and present, namelessness and identity. The infrastructure of the physical organism is at the core of Proust's aesthetics. Without the individual body and its ostensibly irreducible sensory imprints there would be no "genuine" past to recall and, by implication, no story to tell.

Private time versus public time, the temporality of the individual body versus that of the watch: already in the opening pages the narrator creates a dialectical constellation, coupled with a contrast between fantasies of the interior (dreams and reveries) and those of the exterior (the train and the territories it has traversed). Many pages later the narrator will underscore the historicity of temporal consciousness, suggesting that it is profoundly related to changing infrastructures. The railway is now placed at the center of a germinal theory of temporal perception: "Our organs become atrophied or grow stronger or more subtle according as our need of them increases or diminishes. Since railways came into existence, the necessity of not missing trains has taught us to take account of minutes, whereas among the ancient Romans [. . .] the notion not only of minutes but even of fixed hours barely existed" (*REM* 2:853/*RTP* 3:219).

That *Remembrance* is an eloquent argument for the irreducibility of the subjective experience of time is a given. To the extent that the narrative, all three thousand pages of it, knows a climactic denouement, it surely occurs when, in the closing volume, the exalted narrator stumbles on a paving-stone, revisits bygone days in Venice, and so captures "a fragment of time in the pure state" (*REM* 3:905/*RTP* 4:451). The moment is precious, not only because it duplicates a long-forgotten event in the distant past, but also because it offers an unmediated temporal experience. The narrator reclaims a fragment of his individual history and at the same time records the passing of time itself.

The subtle insight in the passage on the relations of infrastructure and the organs of temporal consciousness, however, suggests the presence of a text within the text, a metadiscourse or metacommentary that implants a certain historical reflexivity in Proust's aesthetic enterprise, particularly its ardor for what, in *Remembrance,* is conceived of as the authentic experience of time. In a canonical reading of the text, however, the Proustian celebration of *aisthesis* and the pursuit of authentic time would nevertheless seem to preclude inquiries into their historicity, as if such questions were incongruous to the aesthetic spirit of the novel itself. Indeed, Proust's novel is a modernist site where the incompatibility of art and technology has been most strongly asserted. To be sure, the narrator himself often reproduces the typical opposition of art to technology, of genuine art to mere mass production, and of artistic sensibility to mere instrumental knowledge. Consider an episode in *Swann's Way,* where he reflects upon the beauty of bad weather and dwells on his desire to behold a sublime sea storm. He wants the real thing, however, not some artificial spectacle created by engineers:

> Just as the beautiful sound of her voice, reproduced by itself on the gramophone, would never console one for the loss of one's mother, so a mechanical imitation of a storm would have left me as cold as did the illuminated fountains at the Exhibition. I required also, if the storm was to be absolutely genuine, that the shore from which I watched it should be a natural shore, not an embankment recently constructed by a municipality. Besides, nature, by virtue of all the feelings that it

aroused in me, seemed to me the thing most diametrically opposed to the mechanical inventions of mankind. The less it bore their imprint, the more room it offered for the expansion of my heart. (*REM* 1:417/*RTP* 1:377)

With all the persistence of an educator, the narrator lists not one but three attempts at imitating nature. All fail miserably, and necessarily so, for in the end, nature rejects imitation. Nature is and has to be unique, the narrator suggests, and the same goes for art. Buried within the passage is an entire philosophy of art, one whose conceptual and historical underpinnings I traced in Chapter 1.

But Proust would not be Proust if it were that simple. As we shall see, the humorous diatribe against engineering leads one astray. Anyone who reads Proust's novel carefully will discover how difficult it often is to keep categories of art apart from those of technology. The smell of petrol is as epiphanic as the taste of a madeleine dipped in tea; the airplane above the treetops in Balbec is as sublime as a sea storm. In what follows, I shall argue that the advent of modern technology is constitutively inscribed in Proust's artistic endeavor from the start, and that it makes itself felt on a variety of formal levels. What have been identified as typically Proustian themes—the irreducibility of sensuous experience, the predilection for subjective time, and the interest in individual memory—may usefully be thought together with certain technological innovations in the second machine age, photography and cinematography in particular. Such a juxtaposition, I believe, yields a richer understanding of the Proustian aesthetic.

Proust's novel is concerned, among many other things, with the discovery by the narrator of a visual aesthetics that rests upon the distinction between numbing habit and unmediated sensory experience, between agreed-upon intellectual notions and unbiased forms of comprehension. As the narrator warms to the didactic tasks he has set himself, notably in the last volume, *Time Regained,* he repeatedly criticizes photography and cinematography, declaring them inferior to the human eye, the art of writing, and last but not least, the workings of memory. But things are far more complicated. By discussing the French physiologist Etienne-Jules Marey's work, the philosophical controversies provoked by chronophotography, Henri Bergson's theory of intuition, John Ruskin's celebration of the innocent eye, and popular fairground attractions that simulated speed and movement, I shall demonstrate that photographic modes of representation are fundamental to the success of Proust's visual aesthetics.

I shall pay particular attention to a little-known piece by Proust, "Impressions de route en automobile" from 1907. In this enthusiastic representation of the pleasures of motoring, the car enables a visual experience that explodes habitual modes of seeing. As we shall see, the automobile figures as a cinematic framing device on wheels. The motoring piece was of paramount importance to the novel that Proust began writing the following year.[5] It demonstrates also that the author of *Remembrance* was at once more Ruskinian and more modernistic than has previously been thought, not just in terms of subject matter but in terms of style.

The world Proust sets out to describe in *Remembrance* has already been transformed by the train. No sooner has the dust settled than the automobile makes the winds of history blow up once more. But the railway and the automobile do more than just widen the insuperable gap between past and present. They are part of a modernity that, like photography and film, makes the world writable once again; and it is this very world that Proust hastens to explore before it, too, is devoured by the insatiable appetite for the new. We know how brilliantly Proust pursues times past, and how skillfully he sounds the depths of memory, but the extent of his historical originality becomes apparent only when he places his narrator firmly in the present and has him discover the writability of the modern world.

Optical Overtures: Proust and the Kinetoscope

In the dreamlike overture of the novel, the narrator mentions three optical toys in passing: a kaleidoscope, a kinetoscope, and a magic lantern. Peripheral though they may seem, it is no coincidence that they appear in the opening pages. They situate the novel in a visual order mediated by the immanence of the embodied subject. No transcendental principle ensures the intelligibility of the Proustian world, and the human eye has long since ceased to serve as an emblem of truth and objectivity. In this sense, there are obvious affinities between *The Magic Mountain* and *Remembrance*. In both Mann and Proust, the representation of visual technology emerges as a primary means of thematizing and exploring the disjunction between knowledge and vision. Although both novels revolve around the issue of how knowing relates to seeing, Mann tends to focus on questions of knowledge, whereas Proust tends to tip the scale in favor of questions of seeing. More specifically, where *The Magic Mountain* stages the making of an intellectual and explores epistemological questions, *Remembrance* endeavors to articulate a phenomenology of visual perception, an innocent and unbiased mode of seeing that is at the core of the narrator's aesthetic apprenticeship.

As Roger Shattuck has argued, the language of optics is crucial in *Remembrance*. It is not just a question of frequency; optical metaphors also serve to structure the narrative: "It is principally through the science and the art of *optics* that [Proust] beholds and depicts the world."[6] Indeed, the narrator defines the task of literature by likening the book to a viewing instrument (*REM* 3:949/*RTP* 4:489–90). In similar fashion, he likens his own literary work to a magnifying glass, and his artistic method to a telescope (*REM* 3:1089;1098/*RTP* 4:610;618). These metaphors have a scientific ring, to be sure, but in Proust the truths of art and literature are always those of the individual. Consider the passage when the narrator awakes from his slumber in the opening pages. As he looks around, the dark bedroom appears to him like a kaleidoscope, a metaphor that works to suggest how he fails to identify the flickering interior in which he finds himself (*REM* 1:4/*RTP* 1:4). Flooded with memories of past bedrooms that seem to him strangely real, he is transported to long-forgotten places. He revisits his grandparents' house in Combray, recalling how he had just about learned to endure his melancholic bedroom when a well-meaning person presented him with a magic lantern so that he might amuse himself while waiting for dinner. But the magic lantern only aggravated his apprehension, as the shifting light made him unable to recognize his room.[7]

For the waking mind, such moments of spontaneous retrospection are impossible to grasp one by one, the narrator emphasizes. Always in flux, memory images are as evanescent as they are indivisible. In order to illustrate the essentially mobile nature of memory, the narrator then turns to the image of the *kinétoscope:*

> These shifting and confused gusts of memory never lasted for more than a few seconds; it often happened that, in my brief spell of uncertainty as to where I was, I did not distinguish the various suppositions of which it was composed any more than, when we watch a horse running, we isolate the successive positions of its body as they appear upon a bioscope [*les positions successives que nous montre le kinétoscope*]. (*REM* 1:7/*RTP* 1:7)

The kinetoscope, a stationary motion picture viewer, was developed by Thomas A. Edison in the late 1880s; and according to the dictionary *Le Robert*, the word *kinétoscope* first appeared in the French language in 1893. Inspired by Marey's chronophotographic camera and electric zoetrope,

Edison's peep-show machine became a worldwide success, forced out of business only with the advent of the film projector and the movie hall.[8] But the passage refers to more than the kinetoscope. As William C. Carter has suggested, it also alludes to Eadweard Muybridge's stop-motion photographs of running horses.[9]

In seizing on the image of the kinetoscope, Proust's narrator seeks to contrast the experience of memory to the attempt to analyze its component parts. The flow of memory images, he maintains, is impossible to scrutinize the way one would inspect individual frames on a film strip. The act of remembering and the attempt to analyze this act are as incompatible as the way in which a human eye perceives a trotting horse and the camera's way of registering the same course of events. In short, memory is to analysis as human vision is to the camera eye. As the novel unfolds, the narrator will underscore more than once that memory cannot be understood on the model of photography. Authentic memory is everything that the mechanically produced image is not.

The passage about the kinetoscope is remarkably dense. Brief though it may be, virtually all of Proust's grand themes echo in the sentence: dreaming, memory, time, and perception. At the same time, the passage brings a lively debate to the fore, a debate that had been current for three decades when Proust, in 1908, began writing his novel. Engaging scientists as well as thinkers and artists, the debate revolved around the chronophotographic analysis of bodies in motion. It was one of the most significant philosophical controversies in Proust's time. The human eye was set off against the eye of the machine, in a battle that triggered questions concerning the nature of time, of knowledge, and of objectivity.

Marey, Bergson, and the Crisis of Human Vision

The invisible aspects of bodily movement were Etienne-Jules Marey's prime object of study. Beginning in the 1850s, the French physiologist developed various techniques for exploring the duration and intensity of bodily movement, all measured as a function of time. Marey invented his own instruments, from the sphygmograph, or pulse writer, to the time-lapse camera. They had one thing in common: they translated ephemeral physiological actions into permanent and quantifiable records.

As Marta Braun has shown in *Picturing Time,* Marey's work had a tremendous impact on medical diagnostics and graphic notation, and also on the emergence of cinematography and the so-called European science of labor.[10] The human body is the primordial mediator between these distinct yet interrelated enterprises. The body is the interface that, together with various graphing techniques that encode its physiological actions, creates a theoretical trajectory linking pressure changes in the arterial wall to cardiovascular labor to muscle fatigue to motion pictures to the assembly line, all the way to the work of artists such as Degas, Seurat, and Duchamp. As we shall see, this trajectory also intersects with Proust's *Remembrance*.

In 1860, Marey completed a mechanical device for measuring the pulse. Wired around the wrist, the sphygmograph included a stylus that traced the pressure changes in the arterial wall and inscribed them on a strip of smoke-blackened paper. Marey went on to design a series of instruments that similarly translated physiological actions into graphic representations, such as heat changes, respiration, and muscular contractions. The physiological interior thus became subject to new matrices of legibility.

In Marey's view, the human senses were an imperfect means of discovering objective truths. Not only did he maintain that his instruments were superior to the human observer; he also held that they could lay claim to a perceptual domain all their own: "When the eye ceases to see, the ear to hear, and the sense of touch to feel, or when our senses appear to give us deceptive appearances, these devices are like new senses of an astonishing precision."[11] In addition, he argued that his inscription apparatuses provided the scientist with a precise and unambiguous language.[12]

In the late 1870s, Marey began experimenting with photography. Inspired by Muybridge's studies of galloping horses, he decided that photography was eminently suited to the scientific analysis of how bodies move in space (figure 4). Marey devised a photographic gun with a revolving light-sensitive circular plate. Later, he perfected the traditional camera so that it was capable of multiple exposures on a single plate. In this way, Marey was able to, first, produce precise representations of the moving body at equal temporal intervals; and, second, control and verify them. Like the early graphing instruments, Marey's camera thus functioned as an inscription device. It translated invisible bodily phenomena into visual emblems, and ultimately into graphs (figure 5). Now that sense data could be recorded by technological means, Friedrich Kittler suggests, it had become possible to move "far below the threshold of perception."[13]

Marey's next experiment had its own peculiar logic. He turned the photographic representation of bodily movement into a graphic record proper; he digitized the visual data, as it were. Marey dressed his object of study—always men—in tight black clothes, placed white buttons on each

4. Eadweard Muybridge, *Running (Galloping)*. 1878–1879. Iron salt process. J. Paul Getty Museum, Los Angeles.

5. Etienne-Jules Marey, *Image d'un oiseau*. 1899. Halftone reproduction from *La chronophotographie* (Paris, 1899). J. Paul Getty Museum, Los Angeles.

joint, linked the buttons to one another with metal bands, and then photographed the dark body against a black background. As a result, only the buttons and the metal bands remained visible. The final visual product thus offered a powerful abstraction of the living body, its kinetics henceforth readily disposed to quantification and mathematical analysis.

Marey's work revolutionized physiology. As Bruno Latour has emphasized, chronophotography provided inscriptions of space-time that allowed scientists, particularly physiologists, to venture further into the project of turning aspects of reality into entities fit for calculation and stratification. "The irreversible flow of time was now synoptically *presented* to their eyes. It had in effect become a space on which [. . .] rulers, geometry, and elementary mathematics could be applied."[14] Once the concreteness of the individual body had been turned into a diagram, it was a short step to the science of labor and the concomitant endeavor to rationalize the movements of the worker.

Siegfried Giedion has aptly characterized Marey as a scientist who saw his objects with the "sensibility of a Mallarmé."[15] As early as 1889, the philosopher of art Paul Souriau wrote a treatise, *The Aesthetics of Movement*, in which he encouraged artists to turn to the work of Muybridge and Marey for inspiration. He suggested that the task of the artist is to select the most accurate, expressive, and beautiful chronophotographic images and then base his or her work on this selection. In this way, the artistic representation of movement would include within itself hundreds of other images, and there-

6. Etienne-Jules Marey, *La marche* and *La course rapide*, 1893. From *Etudes de physiologie artistique* (Paris, 1893). J. Paul Getty Museum, Los Angeles.

fore successfully produce in the viewer an immediate impression of motion. A few years later, in 1893, Marey published an artist's book that was to become a tremendous success, *Etudes de physiologie artistique faites au moyen de la chronophotographie* (figure 6). Meanwhile, artists such as Degas and Seurat started using chronophotography as a point of departure in their studies of ballet dancers. Degas also made a wax sculpture of a horse with all four hoofs in the air at the same time, called *Horse Trotting, the Feet Not Touching the Ground*.[16]

Marey's influence on the visual arts should not be underestimated. According to Braun, the impact of Marey's work on the visual arts was probably greater than that of any scientific enterprise since the discovery of perspective in the Renaissance.[17] Avant-garde artists, too, looked to Marey. The most famous example, and surely the most analogically complete counterpart, is Duchamp's *Nude Descending a Staircase* (figure 7). Like Giacomo Balla, František Kupka, Umberto Boccioni, and other artists, Duchamp was intrigued with Marey's work, and he once stressed that chronophotography—not film—inspired the scandalous painting.[18]

Yet, as Braun also observes, chronophotography was far from uncontroversial. Marey's time-pictures rendered scientific truth, to be sure, but, critics asked, did they render artistic truth? Would not a fluid, blurred movement, rather than those analytic freeze-frames, be more truthful to the human experience of time and movement? And how could chronophotography represent objectivity when the eye failed to perceive those frozen segments of time?

In addition to aesthetic debates, chronophotography also stirred philosophical discussions. The philosopher of art Robert de la Sizeranne, in *La photographie, est-elle un art?* (1899), offers an elegant summary of the epistemological issues involved:

> The truth of science is a truth of detail; the truth of art is a truth of ensemble. When chronophotography brings us a print noting one of the thousand phases of which a movement is composed, we respond: That is a part of movement, it is not movement. It is true that one can find the attitude you discovered in a movement, but it is no less true that there are hundreds of others and that *it is the result of all these attitudes—each one immobile during an instant of reason—that forms what we call movement*. My eyes perceive only the ensemble; your camera perceives only a part. Who is to determine who perceives the truth and that my eyes are mistaken?[19]

The philosopher of intuition Henri Bergson, for one, rejected the truth claims of Marey's scientific enterprise. Stressing that the nature of movement is synthesis and flux, and that the camera merely registers frozen moments of time, Bergson argued that chronophotography adds nothing to the understanding of movement. Bergson's critique of chronophotography is far from peripheral to his philosophy. According to recent scholarship, significant parts of Bergson's writings on the perception of time,

7. Marcel Duchamp, *Nude Descending a Staircase, no. 2*, 1913. Philadelphia
Museum of Art, the Louise and Walter Arensberg Collection. © 2001 Artists
Rights Society (ARS), New York / ADAGP, Paris / Estate of Marcel
Duchamp.

in particular *Matter and Memory* (1896), *Introduction to Metaphysics* (1903), and *Creative Evolution* (1907), are to be understood as a commentary on the theoretical and cultural implications of Marey's inventions, although Bergson never mentioned him by name.[20]

In the widely influential *Creative Evolution,* chronophotography helps Bergson define what he thinks of as the genuine perception of time and movement. The concluding chapter, entitled "Le mécanisme cinématographique de la pensée et l'illusion mécanistique," is partly based on Bergson's lectures at the Collège de France during the years—1900 through 1904—when Marey's tenure overlapped with his. During this period, he lectured on the history of the concept of time; and as he explains in a footnote, he then compared the mechanism of conceptual thought to that of the *cinématographe*—a comparison he was to carry over into *Creative Evolution.*

In choosing cinematography as his central metaphor, Bergson seeks to make visible the essence of conceptual thought: analysis. He does not seize on what is typically associated with cinematography—that is, movement. He dwells instead on the constitutive element of cinematic movement: the decomposition of the real into so many distinct images that are then put back together by the mind and, because of retinal retention, perceived as movement. That Bergson associates cinematography with decomposition is easier to understand once one realizes that very little separates Marey's projecting apparatuses from those developed by the Lumière brothers. The defining moment of commercial cinematography is generally attributed to Auguste and Louis Lumière who, initially, had aligned their interests with those of Marey, and who successfully patented their "projecting chronophotographe" in early 1895, a camera that shared many features with Marey's apparatuses. Soon thereafter they refined the device somewhat, patented it once again, now under the name *cinématographe,* and then exhibited their motion pictures to an audience for the first time.[21]

Bergson's point is that once the camera has translated the real into images, there is always a loss. The experience of the whole is therefore an illusion. This essentially analytic mechanism—the selection that comes to represent the whole—provides the model on which he understands conceptual or scientific thought as such. Historically speaking, Bergson maintains, this loss had little significance, as the sciences in the past were concerned with qualities. Modern science, however, is directed at quantities. From Galilei onward, physics is characterized by its "indefinite breaking up of time," the ultimate purpose being not to understand essences (quality) but to be able to measure (quantity).[22] This is why modern science is fundamentally unable to grasp the innermost core of being: flux.

It is at this point in the argument that Bergson mobilizes the example of chronophotography. He turns Marey's work into a metaphor of modern science at large. Seeking to bring out the difference between quality and quantity, Bergson maintains that Greek physics relates to modern physics as ocular perception of movement relates to photographic means of registering movement. It is no doubt true, he argues, that the camera offers a more exhaustive representation of how a body moves in space, but the selection is arbitrary and has nothing in common with human perception:

> Of the gallop of a horse our eye perceives chiefly a characteristic, essential or rather schematic attitude, a form that appears to radiate over a whole period and so fill up the time of gallop. It is this attitude that sculpture has fixed on the frieze of the Parthenon. But instantaneous photographs [*la photographie instantanée*] isolate any moment; it puts them all in the same rank, and thus the gallop of a horse spreads out for it into as many successive attitudes as it wishes, instead of massing itself into a single attitude, which is supposed to flash out in a privileged moment and illuminate a whole period.[23]

This remarkable passage, like Proust's kinetoscope passage, clearly refers to chronophotographic analyses of motion, enterprises for which the horse frozen in midair served as an emblem. We also glimpse some of the vital distinctions through which Bergson's philosophy of intuition is framed: quality and quantity, synthesis and analysis, the human and the mechanical, the eye and the machine, ancient Greece and modern times, and visual art and photography.

Bergson thus turns Marey's project on its head. Or, rather: Bergson and Marey agree on the theoretical distinctions but disagree on the distribution of value. The former attaches a positive value to the terms to which the latter attaches a negative one, and vice versa. This is to say that Bergson and Marey, concerned as both of them are with the epistemological status of the human eye and the representation of time and movement, operate within one and the same theoretical paradigm.

Marey's innovations, from the sphygmograph to chronophotography, constitute a key moment in the deperceptualization of the natural sciences. Demonstrating that seeing was not necessarily knowing, they helped establish a discursive division of labor that, more radically than ever before, separated the epistemic from the sensory, and the scientific from the artistic. The human eye turned out to have very little to do with *theoria* and all the more with *aisthesis*. Meanwhile Marey's inventions functioned, as he himself put it, like new senses.[24]

On the face of it, Proust would seem to agree with Bergson's philosophical critique of chronophotography. After all, Proust, too, rejects the camera as a means of picturing time. He, too, celebrates heterogenous temporality. He, too, embarks on an inquiry into memory as a bodily process mediated by the human sensorium. No wonder, then, that there is a striking affinity between Proust's sentence about the kinetoscope and Bergson's passage about instantaneous photography. Indeed, both use a chronophotographic image of a horse in motion to define the nature of genuine perception; and both create a double analogy to bring home their points—Proust making visible the irreducible gap between the fluidity of authentic memory and the attempt to analyze its very nature, Bergson making visible the discrepancy between a mode of thought directed at qualities and one directed at quantities. It is therefore all the more interesting that, as Julia Kristeva has observed, Proust wrote not *kinétoscope* but *cinématographe* in his first drafts of the passage.[25]

For all these affinities, however, Proust has little in common with Bergson. A startling complexity marks the aesthetics of perception for which *Remembrance* becomes a vehicle. There is a tremendously intricate relationship between, on the one hand, Proust's ideal of unmediated perception and, on the other hand, the deployment of photographic techniques of representing visual experience. For not only does Proust appropriate such modes of representation; he puts them to use at precisely those moments where he attempts to elaborate what one may call a phenomenology of perception, that is, when he seeks to render perceptual activities that are uninhibited by intellectual preconceptions. What is more, some of Proust's boldest stylistic solutions owe their deepest impulses to machines of vision. Indeed, photographic means of representation, including chronophotographic and cinematographic ones, are singularly well suited to Proust's ambition to restore sensory impressions that have not been domesticated by the dulling force of habit.

Before turning to Proust's aesthetics of perception, I want to add a point to the discussion of Marey. The advent of chronophotography helped initiate a new discourse of the human body. Increasingly, the physical organism was seen as a laboring unit, conceived of as a machination of bio-forces functioning on the model of mechanics and thermodynamics. This conception then came in tandem with its dialectical counterpart: the idling body, the body of *aisthesis* and of noninstrumental sensory pleasure. In *Remembrance*, as I have already suggested, it is the human body that founds the Proustian aesthetic. After all, Proust's theory of memory is indistinguishable from his aesthetics of perception. If, then, the body is the occasion for the aesthetics of memory for which *Remembrance* serves as a vehicle, the body implied is emphatically individual, singular, and irreducible. Yet this too needs qualification; the body implied is above all a sensorium, an ensemble of refined sensibilities and compartmentalized sensory experiences. Such a view of the body, as the following pages will suggest, is a dialectical mirror image of Marey's universalized notion of the laboring body. Both constructs derive from the same discursive system, and both, crucially, have been inscribed in the universe of Proust's novel, not so much as a contradiction as two competing yet mutually dependent and intertwined impulses, eminently suited to Proust's aesthetic pursuits.

The Education of the Eye

An episode in *Within a Budding Grove* takes us to the heart of the novel's visual aesthetics. The narrator is vacationing in Balbec, a seaside resort in Normandy. One hot summer day, the narrator pays a visit to the painter Elstir. A fictitious figure, Elstir is a composite of an array of nineteenth-century painters, first of all Turner, but also artists such as Monet, Whistler, Moreau, and Vuillard; in addition, Elstir is designed to remind the reader of Ruskin. No sooner has the narrator entered the artist's studio than he is seized by happiness and the prospect of rising to a poetical understanding of reality. Elstir's atelier, he proclaims enthusiastically, is nothing less than a "laboratory of a sort of new creation of the world" (*REM* 1:892/*RTP* 2:190). The canvases lining the studio, most of them seascapes, make him perceive the sea anew, and the beach and the sky and the foaming waves. The experience elicits a piece of art criticism, as spirited as it is ingenious:

> [Elstir] had extracted, by painting them on various rectangles of canvas that were placed at all angles, here a sea-wave angrily crashing its lilac foam on to the sand, there a young man in white linen leaning on the rail of a ship. The young man's jacket and the splashing wave had acquired a new dignity from the fact that they continued to exist, even though they were deprived of those qualities in which they might be supposed to consist, the wave being no longer able to wet or the jacket to clothe anyone. (*REM* 1:892–93/*RTP* 2:190)

Elstir has freed the wave and the jacket from their established denotations, thereby restoring the meanings that habitual perception prevents the beholder from perceiving. No longer subject to a reality that revolves around usage and function, the wave and the jacket have entered a world of purposeless beauty. In this way, the narrator intimates that Elstir is not a realist, much less a naturalist; if anything, the painter is an impressionist—he is faithful to the visual impression and the immediate ways in which it comes before him. He transforms the visible world, at the same time restoring its lost glory; and the beholder is taken back to the dawn of creation:

> If God the Father had created things by naming them, it was by taking away their names or giving them other names that Elstir created them anew. The names which designate things correspond invariably to an intellectual notion [*notion de l'intelligence*], alien to our true impressions, and compelling us to eliminate from them everything that is not in keeping with that notion. (*REM* 1:893/*RTP* 2:191)

For a few blessed moments, the world of phenomena appears as though it had just been born. The conventional links between signifier and signified have been severed. Elstir's prime means is the painterly equivalent of the metaphor—the metamorphosis. This is why the narrator seizes on those seascapes where the contours of land and sea blend and eventually change places, where the town rising above the sea has been rendered in "marine terms," and the sea has been depicted in "urban terms." To be sure, the beholder will eventually recognize the sea and the town, but through notions not typically associated with them. This is the sense in which Elstir not only strips things of their common names but also, and above all, endows them with new ones. And in naming things anew, the artist invites the beholder to see them anew, richer, fuller, and more dignified.

The episode also contains an enigmatic description of the studio itself. The blinds are closed, the atelier plunged in half-light. But there is one place, the narrator underscores, "where daylight laid against the wall its brilliant but fleeting decoration, dark; one small rectangular window alone was open" (*REM* 1:893/*RTP* 2:191). A darkened chamber, with a small opening providing the only source of light: Elstir's studio is literally and figuratively represented as a camera obscura, complete with an aperture. The dark room turns into a metaphor for the artist's power to recreate the world as he pleases.[26]

The painter's eye as camera obscura? The metaphor acquires a peculiar resonance, as the episode also includes a lengthy reflection upon the rivalry between painting and photography. The competition with the mechanically produced image has compelled Elstir to reinvent his painterly aesthetics. This is because photography—*une industrie,* Proust writes—has banalized traditional art. "Since Elstir began to paint, we have grown familiar with what are called 'wonderful' [*admirable*] photographs of scenery and towns" (*REM* 1:896/*RTP* 2:194). The appeal of photography lies in its ability to explode habitual ways of seeing by estranging seemingly familiar visual domains, at the same time being a faithful recording of the motif in question: "If we press for a definition of what [the] admirers mean by the epithet [i.e., *admirable*], we shall find that it is generally applied to some unusual image of a familiar object, an image different from those that we are accustomed to see, unusual and yet true to nature, and for that reason doubly striking because it surprises us, takes us out of our cocoon of habit, and at the same time brings us back to ourselves by recalling to us an earlier impression" (*REM* 1:896–97/*RTP* 2:194).

No matter how pleasing and revelatory photographic means of representation may be, and no matter how effectively it ruptures habitual modes of perception, painting, not photography, was first in the field. Historically speaking, the distinction of painting stems from its ability to defamiliarize the familiar and to make to beholder look at the world anew, but the mechanically produced image has appropriated this very domain, with the effect that traditional painting appears obsolete, even lacking in originality. So it is that the sheer existence of photography forces the art of painting to reinvent itself. In other words, in suggesting that the superior ability of photography to represent landscape and urban sceneries has left Elstir with no option but to abandon his classical motifs, the narrator offers a historical explanation of the painter's artistic choices. As Anne Henry has underscored, this is all the more remarkable since Proust, throughout *Remembrance,* tends to deny that history may serve as a valid point of departure for explication in the realm of art.[27] Here, however, history is evoked—in order to explain how Elstir's older mode of painting began to lose legitimacy and how, as a result, he tries to attain an innocent, unbiased, and nontheoretical way of seeing. In short, Elstir comes to celebrate the disinterested spectator—like Turner did in his late period and as Monet and other impressionists were to do later in the nineteenth century. Nathalie Sarraute has captured this sea change in an apt phrase: "Since Impressionism," she writes, "all pictures have been painted in the first person."[28]

Elstir thus gives priority to the visual impression, not the intellect. This is why he always attempts to rid himself of all knowledge before setting to work. That Elstir seeks to remain faithful to the purely visual aspects of the seen does not mean, however, that his paintings are necessarily "true" in the ordinary sense of the word; he may well depict an optical illusion and therefore commit an "error," as when the sea emerges like a town. But the narrator conceives of such illusions as *prior* to the work of the intellect and therefore artistically true (*REM* 1:897/*RTP* 2:194). For all the immediacy of such optical errors, however, the artist nonetheless has to *unlearn* his mode of seeing in order to be able to see what he actually perceives; in fact, he has to accustom his eyes to not recognizing, for example, that fixed boundary that separates land from sea and so tells the beholder to identify the town as a town and the waves as waves (*REM* 1:895/*RTP* 2:192–93). In other words, Elstir's aesthetics of perceptual innocence presupposes the education of the eye.

Turner: Knowing and Seeing

The Elstir episode recalls a famous anecdote about the British painter J. M. W. Turner. Having drawn a harbor seen against the light, Turner shows his work to a naval officer who, somewhat puzzled, remarks that the ships lack portholes. "No, certainly not," Turner replies. "If you will walk up to Mount Edgecumbe, and look at the ships against the sunset, you will find you can't see the port-holes." The naval officer retorts: "Well, but you know the port-holes are there." "Yes," says Turner, "I know that well enough; but my business is to draw what I see, and not what I know is there."[29] Where the painter's eye beholds sun-drenched vessels, the naval officer perceives armed warships. Aesthetic experience and practical reason do not inhabit the same world. To observe an object from a user's point of view, the anecdote implies, is to engage in a perceptual activity radically different from that of a spectator who has no interest in the object other than to, say, paint it.

John Ruskin, who championed Turner's late manner, quoted the anecdote in his treatise on the relations of art and the natural sciences. For the British art critic, the story epitomized an aesthetic program, whose ultimate task was to correct the detrimental effects of industrial society on human experience. As I shall discuss, Ruskin resented many things modern: railway travel, for example, fatigued the human sensorium; photography risked making the eye too fastidious; and viewing instruments such as the Claude glass perverted artistic seeing.

Proust, too, seized on the anecdote. A sincere admirer of Ruskin's philosophy of art, he related it in the preface to his translation of Ruskin's *The Bible of Amiens*.[30] Proust was to make the spirit of the anecdote into his own. Indeed, as Jean Autret has emphasized, it provides a key to the artistic sensibility for which *Remembrance of Things Past* becomes a vehicle and that the portrait of the painter Elstir in particular serves to articulate.[31] Fresh sense perception versus confining intellectual notions: this distinction is fundamental to Proust's visual aesthetics. Teaching the virtues of the cultivation of sensory experience, *Remembrance* articulates a didactics of disinterested perception—of seeing for seeing's sake, of listening for the pleasure of sounds, and so on. The task of the artist is to be alert to the immediacy of sensory experience and, as the narrator underscores in the Elstir episode, to restore it by inventing the appropriate metaphor.

Learning the Lesson: Framed Visions

The encounter with Elstir is a key episode.[32] The narrator alludes several times to the visit, thus making sure that no reader will miss the significance of the crucial insights he takes to heart. In *Within a Budding Grove,* he anticipates how he will come to understand why Mme de Sévigné's letters bring him such pleasure, for she "is a great artist of the same school as a painter whom I was to meet at Balbec, where his influence on my way of seeing things was immense. I realised at Balbec that it was in the same way as he that she presented things to her readers, in the order of our perception of them, instead of first explaining them in relation to their several causes" (*REM* 1:703/*RTP* 2:14).

In *The Guermantes Way,* moreover, the narrator contemplates Elstir's aesthetics with a completeness of detail that recalls his reflections during the studio visit. Invited to a dinner party at the home of the duke and duchess of Guermantes, he asks to see their Elstir collection and forgets about the social gathering:

> Among these pictures, some of those that seemed most absurd to people in fashionable society interested me more than the rest because they recreated those optical illusions which prove to us that we should never succeed in identifying objects if we did not bring some process of reasoning to bear on them. How often, when driving, do we not come upon a bright street beginning a few feet away from us, when what we have actually before our eyes is merely a patch of wall glar-

ingly lit which has given us the mirage of depth. This being the case, it is surely logical [...] to return to the very root of the impression, to represent one thing by that other for which, in the flash of a first illusion, we mistook it. [...] Elstir sought to wrest from what he had just felt what he already knew; he had often been at pains to break up that medley of impressions which we call vision. (*REM* 2:435/*RTP* 2:712–13)

Later in the volume, the narrator makes a comparison between Elstir and himself. The narrator, clearly, is well on his way to incorporating the painter's aesthetics into his own artistic sensibility. "My imagination, like Elstir engaged upon rendering some effect of perspective without reference to the notions of physics which he might quite well possess, depicted for me not what I knew but what it saw" (*REM* 2:590/*RTP* 2:856).

Yet although the Elstir episode relates a crucial discovery, Proust has carefully prepared the way. The manner in which Proust describes the long prelude to the visit—the train ride to Balbec, the atmosphere at the seaside resort, Robert de Saint-Loup's appearance, and so on—is itself exemplary of the aesthetics that the narrator discovers in the studio and eventually will make his own. At this stage in the novel, the stylistic solutions become bolder, as if to reinforce the point that the narrator's artistic education is under way. The hero's surroundings are repeatedly represented as though they were works of art; sometimes they have even been furnished with a frame. Moreover, each time the visual impression takes center stage.

Consider the description of the railway journey which works to warm up the reader. Early in the morning, as the train makes its way toward Balbec, the narrator studies the break of dawn through the square of the window (*le carreau de la fenêtre*). Just above a dark silhouette, he distinguishes "some ragged clouds whose fleecy edges were of a fixed, dead pink, not liable to change, like the colour that dyes the feathers of a wing that has assimilated it or a pastel on which it has been deposited by the artist's whim" (*REM* 1:704/*RTP* 2:15). No sooner has the narrator clapped his eyes on the pink light of dawn than he elevates the sight to a work of art executed by a willful artist. Moments later, daylight bursts forth, and a cascade of colors appears before him:

the sky turned to a glowing pink which I strove, glueing my eyes to the window [*la vitre*], to see more clearly, for I felt that it was related somehow to the most intimate life of Nature, but, the course of the line altering, the train turned, the morning scene gave place in the frame of the window [*le cadre de la fenêtre*] to a nocturnal village, its roofs still blue with moonlight, its pond encrusted with the opalescent sheen of night, beneath a firmament still spangled with all its stars, and I was lamenting the loss of my strip of pink sky when I caught sight of it anew, but red this time, in the opposite window which it left at a second bend in the line; so that I spent my time running from one window to the other to reassemble, to collect on a single canvas the intermittent, antipodean fragments of my fine, scarlet, ever-changing morning, and to obtain a comprehensive view and a continuous picture of it [*une vue totale et un tableau continu*]. (*REM* 1:704–5/*RTP* 2:15–16)

Framed by the windows, the view of the sleepy landscape through which the train travels is literally represented as a series of tableaux in motion.[33] Scenery follows upon scenery, as in a film montage. Such panoramic perception, as Wolfgang Schivelbusch suggests, is inherent in the railway journey. As such, it represents a break in the history of visual perception: "Panoramic perception, in contrast to traditional perception, no longer belonged to the same space as the perceived: the traveler saw the objects, landscapes, etc. *through* the apparatus which moved him through the world. That machine and the motion it created became integrated into his visual perception."[34] Schivelbusch does not state it, but such a visual experience anticipates cinema.[35] In Proust, at any rate, the railway compartment moves through the landscape like a projector apparatus. Naturalizing the deliciously fragmented representation of the narrator's visual activity, the window frames permit the spectacle to come into being. In short, the train emerges as a framing device on wheels.

As the train stops at a little station, and as the narrator catches sight of a peasant girl selling coffee, it becomes clear that the window gazing has aestheticized his way of seeing to such a degree that she immediately turns into a spectacle: "Flushed with the glow of morning her face was rosier than the sky" (*REM* 1:705/*RTP* 2:16). The train ride, the narrator explains, has exposed him to a form of beauty that falls outside of traditional notions and, for this very reason, is proof of genuine beauty. In order to take a good look at the peasant girl, he signals that he wants to purchase a cup of coffee. As she approaches, his gaze frames her appearance and puts it behind glass: "Above her tall figure, the complexion of her face was so burnished and so glowing that it was as if one were seeing her through a lighted window [*vitrail illuminé*]. She retraced her steps. I could not take my eyes from her face which grew larger as she approached, like a sun which it was somehow possible to stare at and which was coming nearer and nearer, letting itself be seen at close quarters, dazzling you with its blaze of red and gold" (*REM* 1:706–7/*RTP* 2:17–18).

The description is remarkable not for what is says about the girl's looks—after all, we learn virtually nothing about her—but for what it reveals about the hero's eye. Elstir would have been pleased. The moment the narrator fixes his gaze on the young woman, her face turns into a light show. He is faithful to what he sees, not to what he knows, and what he sees possesses all the wondrous powers of the idiosyncratic artist's eye. Accordingly, the narrator states not that the young woman's face is coming nearer but that it is becoming larger: "*je ne pouvais détacher mes yeux de son visage de plus en plus large.*" It is as though we were watching a film in which the heroine's head approaches the camera and eventually fills the screen. The peasant girl has become pure image.

Why does the narrator remark that it is as though he looked at her through a lighted window, "*un vitrail illuminé*"? At first, the metaphor may seem like mere embellishment. But Proust needs the image of the window; it motivates the narrator's framed vision and naturalizes his aestheticized perspective. Proust makes use of a similar technique when, a few pages later, he introduces a key character in the novel, Robert de Saint-Loup. The narrator is sitting in the hotel dining-room when he catches sight of the young marquis. It is a scorching hot afternoon, the dining room is "plunged in semi-darkness," and through the gaps between the drawn curtains the glittering blue sea can be glimpsed. Like Elstir's studio, the dining room thus emerges as a camera obscura, complete with apertures. In this giant dark room, the writer-to-be is busy registering remarkable sights, developing one after the other with extraordinary artistic skill:

> [Robert de Saint-Loup] strode rapidly across the whole width of the hotel, seeming to be in pursuit of his monocle, which kept darting away in front of him like a butterfly. He was coming from the beach, and the sea which filled the lower half of the glass front of the hall made a background against which he stood out full-length, as in certain portraits whose painters attempt, without in any way falsifying the most accurate observation of contemporary life, but by choosing for their sitter an appropriate setting [. . .] to furnish a modern equivalent of those canvases on which the old masters used to present the human figure in the foreground of a landscape. (*REM* 1:783–84/*RTP* 2:89)

As in the peasant-girl episode, the snapshot of Saint-Loup is framed by the windows that motivate the delimitation of the narrator's field of vision.[36] This time the scene is explicitly rendered as a work of art. The same is true when the narrator enters his hotel room and turns his eyes toward the window. As in the railway episode, the landscape is transformed into a series of tableaux, complete with frames:

> Gradually, as the season advanced, the picture [*tableau*] that I found there in my window changed. At first it was broad daylight, and dark only if the weather was bad: and then, in the greenish glass which it distended with the curve of its rounded waves, the sea, set between the iron uprights of my casement window like a piece of stained glass in its leads, ravelled out over all the deep rocky border of the bay little plumed triangles of motionless foam. (*REM* 1:860/*RTP* 2:160)

The narrator hastens to add that the foaming sea recalls Pisanello's drawings of feathers and down, and that the creamy white background makes him think of Emile Gallé's etched glass objects. Waves that look like feathers, foam that appears as down: such a way of seeing is perfectly consistent with Elstir's program. The task of the artist is to remain faithful to the seen, to give priority to the imagination over and above confining intellectual notions, and the art-historical figure who in this respect exercised the single most important influence on Proust was John Ruskin.

Ruskin and the Innocence of the Eye

Proust's interest in Ruskin, the most influential art critic of the Victorian age, began in 1899.[37] During a vacation in the mountains, he read Robert de la Sizeranne's *Ruskin et la religion de la beauté* (1897). Interestingly, as Richard Macksey has noted, Proust first invoked Ruskin as an "instrumentality of sight," since he wanted to read Sizeranne's study of Ruskin in order that he, as he said, might "look at the mountains with the eyes of this great man."[38] On the occasion of Ruskin's death in 1900, Proust wrote an obituary notice in *Le Figaro*, suggesting that the proper way to pay homage to the British art critic was to visit the French cathedral cities he had written about, as these now "protect his soul."[39] Proust himself was to set out on Ruskinian pilgrimages, particularly in Normandy, but he also went as far as Italy with Ruskin's writings as a guide. Most of these trips have left traces in *Remembrance*.

A passionate advocate of Ruskin's theory of art, Proust translated two of his books into French, *The Bible of Amiens* (*La bible d'Amiens,* 1904) and *Sesame and Lilies* (*Sésame et les lys,* 1906), both of which he provided with voluminous prefaces and annotations. What attracted Proust to Ruskin was, among other things, the emphatic distinction between the observed and the known, between fresh sensation and dulling habit. "The greatest thing a human soul ever does in this world is to *see* something, and tell what it *saw* in a plain way," Ruskin writes in his seminal work on Turner's late manner, *Modern Painters.* "To see clearly is poetry, prophecy, and religion,—all in one" (*WJR* 5:333; italics in the original). Stressing that the artist must recover the "innocence of the eye," Ruskin promoted "a sort of childish perception of these flat stains of colour, merely as such, without consciousness of what they signify,—as a blind man would see them if suddenly gifted with sight" (*WJR* 15:27n).

Ruskin's hostility to the modernization process runs through all his writings, a fact, incidentally, that did not escape Proust.[40] Ruskin railed against the "current political economy," including the automation of production and the mechanization of labor; and these sentiments were integral to his theory of the visual arts. When Ruskin championed Turner's art, he argued for the painter's superiority by setting up an opposition between the mechanical and natural or, alternatively, the mechanical and the spiritual. Contrasting Turner with old masters such as Lorrain, Rosa, and Poussin, Ruskin thus claimed that they submitted to mechanical, hence false, reproductions of nature, as they failed to rise above "the daguerreotype or calotype, or any other mechanical means that ever have been or may be invented" (*WJR* 3:169). Therefore, Ruskin concluded, they lacked love for the phenomena represented.

As Robert Hewison has underscored, Lorrain, Rosa, and Poussin were associated with the so-called picturesque.[41] Deriving from the Italian word *pittoresco*, or "in the manner of painters," the specifically British notion of the picturesque complemented Burke's distinction between the sublime and the beautiful. Eminently suited to the English countryside, the picturesque referred to the particular pleasure evoked by natural phenomena that were irregular, rough, and wild—without, however, inspiring awe in the spectator.[42] The picturesque taught spectators how to perceive landscape by way of paintings. In other words, it cultivated not pure but derived ways of seeing.

The so-called Claude glass was part of the didactics of the picturesque. A pocket-sized black piece of convex mirror-glass, the instrument framed and reflected the surrounding landscape, thus turning the view into something like a picturesque work of art. Ruskin's critique of such visual pleasures was relentless. The glass, he argued in *The Elements of Drawing*, is "one of the most pestilent inventions for falsifying Nature and degrading art which ever was put into an artist's hand" (*WJR* 15:201–2).

In a similar spirit, Ruskin attacked photography and its automatisms in contrast to the inherent virtues of painting.[43] Artistic seeing was for Ruskin a subjective affair, its empire defined by the boundaries of industrial society. For observation to quicken, and for sensation to sharpen, he recommended as "little change as possible." Ruskin evoked a rural Arcadia lingering in the shadow of industrial society:

> If the attention is awake, and the feelings in proper train, a turn of a country road, with a cottage beside it, which we have not seen before, is as much as we need for refreshment; if we hurry past it, and take two cottages at a time, it is already too much: hence, to any person who has all his senses about him, a quiet walk along not more than ten or twelve miles of road a day, is the most amusing of all travelling; and all travelling becomes dull in exact proportion to its rapidity. Going by railroad I do not consider as travelling at all; it is merely "being sent" to a place, and very little different from becoming a parcel. (*WJR* 3:370)[44]

Proust valued Ruskin's sensibility highly. He adopted Ruskin's belief in the primacy of sense perception and of the importance of remaining true to the concrete. He made use of Ruskin's idea that the sensuous world is a universe of signs that the artist is meant to decipher. Yet, for all his admiration for the British art critic during the critical years that preceded his great novel, Proust was nevertheless to invert Ruskin's attitude toward the historical conditions that had prompted some of the most important features of the latter's aesthetic program.

The Hierarchy of Arts

One of the thematic threads that runs all through Proust's tale is the narrator's fear that he might not become what he so desperately wants to be—an artist. Year in and year out, the narrator is burdened by his utter incompetence, until he finally stumbles on that famous paving-stone and realizes that he will write the very novel that the reader is about to finish. Incorporating its own beginnings, Proust's narrative thus renders its own coming into being. But *Remembrance* is more than a metanovel. As the Elstir episode suggests, it also articulates an aesthetic theory, a theory for which the novel itself serves as an example. It makes itself felt in several ways, explicitly in the closing volume, *Time Regained;* implicitly in episodes on various artists and their achievements, from Berma and Vinteuil to the Goncourt brothers. This aesthetic theory, furthermore, comes with an implied hierarchy of cultural forms. Architecture, painting, music, drama, and literature are all valorized. By virtue of their unique presence in time and space, they figure, as Benjamin would have argued, as auratic modes of cultural production. It might be objected that literature necessarily deviates from this scheme, since its distribution in time and space is vitally dependent on printed copies. This, however, does not bother Proust's narrator. Indeed, as Victor Graham has stressed, for Proust as for the narrator, literature is a spiritual entity.[45]

Photography and film, on the other hand, are denigrated throughout *Remembrance*. In Proust, photography and film are seen as deficient techniques for mnemonic storage and visual mimesis. They are counterposed to human memory and the art of representation, whether visual or verbal. Of course, in Proust's time such a view was common enough. What should surprise us is that photography and film appear in the novel in the first place and, furthermore, that they appear in key episodes. When, for example, the narrator looks at a photograph of his grandmother, it makes him unable to remember her, thereby blocking his suffering; and suffering is what connects him to her. It is as though the print makes him lose his beloved grandmother once more (*REM* 2:786/*RTP* 3:156–57). Only when the narrator unties his shoes does her memory make itself felt. He recalls how she, although aged and tired, used to care for him, bending down to untie his boots. For the first time, a year after her death, her living memory thus comes before him, unexpectedly flashing into the present. As Susan Sontag has observed, whenever Proust mentions photographs, "he does so disparagingly: as a synonym for a shallow, too exclusively visual, merely voluntary relation to the past, whose yield is insignificant compared with the deep discoveries to be made by responding to cues given by all the senses—the technique he called 'involuntary memory.' "[46] In Proust, photography thus comes to represent everything that genuine memory is not. Mechanical, voluntary, and nonselective, photography merely offers a static form of knowledge of human life.[47]

The narrator also asserts that the eye functions in ways far distant from those of a camera. At a social gathering, the duchess of Guermantes claims that the seventeenth-century painter Frans Hals is so magnificent that a mere glimpse of his work—through the windows of a tram, for example—is all it takes to prove its sublime character. The idea startles the narrator: "This remark shocked me as indicating a misconception of the way in which artistic impressions are formed in our minds, and because it seemed to imply that our eye is in that case simply a recording machine which takes snapshots [*instantanés*]" (*REM* 2:544/*RTP* 2:813).

Despite such criticisms, however, the narrator cannot refrain from drawing on the vocabulary of photography, not even when he intends to underscore the incompatibility between human memory and photography: "I tried next to draw from my memory other 'snapshots,' those in particular which it had taken in Venice, but the mere word 'snapshot' made Venice seem to me as boring as an exhibition of photographs, and I felt that I had no more taste, no more talent for describing now what I had seen in the past, than I had had yesterday for describing what at that very moment I was, with a meticulous and melancholy eye, actually observing" (*REM* 3:897–98/*RTP* 4:444).

What is striking is not so much the narrator's proposition that memory cannot be understood on the model of photography, but that memory, in *Remembrance,* is so frequently placed in contradistinction to photography. Put differently, the narrator needs photography for his theory of memory; photography is mobilized precisely for us to understand the true nature of memory—by way of that which it is not. The same is true of cinematography; it, too, is posited as alien to mnemonic processes. Unlike photography, however, film is absent from the plot. Proust's narrator goes for automobile rides, is fascinated by bicycles, pores over photographs, dispatches telegrams, reads newspapers, compares his gaze to an X-ray camera, often talks on the phone, and even fantasizes about a photo-telephone, but he never watches film. As Shattuck has pointed out, "it will remain a mystery how Proust missed exploiting the magic name of Lumière."[48] Indeed, given the impressive number of significant scientists, inventors, and innovators mentioned in *Remembrance,* this silence is remarkable.

But even if movie-watching is absent from the plot, and even if the word "cinematography" appears only a few times, film nevertheless occupies an important place in Proust's novel. Film helps the narrator define his theory of art. In the aesthetic discussions in *Time Regained,* where the narrator is at his most didactic, he underscores several times that cinematography has little to do with the art of the novel: "Some critics now liked to regard the novel as a sort of procession of things upon the screen of a cinematograph. This comparison was absurd. Nothing is further from what we have really perceived than the vision that the cinematograph presents" (*REM* 3:917/*RTP* 4:461).

Such an implicit theory of value is all the more interesting since Proust's metanovel tells the story of the making of a man of letters. Vigorously opposing cinematography to the art of the writer, the narrator goes on to demonstrate the affinity between literature and reality. Consider how time acquires depth and how the real gradually releases itself from stubborn generalities:

An hour is not merely an hour, it is a vase full of scents and sounds and projects and climates, and what we call reality is a certain connexion between these immediate sensations and the memories which envelop us simultaneously with them—a connexion that is suppressed in a simple cinematographic vision, which just because it professes to confine itself to the truth in fact departs widely from it—a unique connexion which the writer has to rediscover in order to link for ever in his phrase the two sets of phenomena which reality joins together. (*REM* 3:924/*RTP* 4:467–68)

The passage plays off the mere measuring of time, those empty and meaningless quantities, against *l'expérience vécue* and the irreducible sensations and memories that constitute lived experience. It establishes a distinction between homogeneous and heterogeneous time, quantity and quality, that echoes widely in modernist art. Literature and cinematography, polemically counterpoised, thus emerge as rivals in the struggle for claims to truth and reality:

If reality were indeed a sort of waste product of experience, more or less identical for each one of us, since when we speak of bad weather, a war, a taxi rank, a brightly lit restaurant, a garden full of flowers, everybody knows what we mean, if reality were no more than this, no doubt a sort of cinematograph film of these things would be sufficient and the "style," the "literature" that departed from the simple data that they provide would be superfluous and artificial. But was it true that reality was no more than this? (*REM* 3:925/*RTP* 4:468)

But of course the narrator will conclude that reality is more than this; the real resides in that supposedly more authentic interior that comes into existence at the juncture of sense impressions and the memories elicited thereby. This is where Proust's sensory epiphanies enter the theory. The tasting of the madeleine biscuit, the sight of the hawthorn hedge, the sound of a spoon against a saucer, the unevenness of two paving-stones: they are the catalysts for the world evoked by *Remembrance.* As the past collides with the present, an inner being is temporalized and brought to life, a supposedly more authentic being that draws its sustenance from what the narrator calls the essence of things and from them alone. And nothing, as the narrator maintains yet again, could be further from the real than cinematography.

What is significant, then, is not so much that media of cultural production such as photography and cinematography are excluded from the notion of authentic art and culture proposed by *Remembrance,* but rather that they operate as scaffolding for the narrator's propositions concerning great art. We have seen how the narrator alludes to the emergent crisis of art triggered by the advent of new media, such as photography and cinematography. The Elstir episode thus serves as a metacommentary on *Remembrance* and some of its most characteristic aesthetic solutions. In other words, Proust's novel incorporates, on the level of enunciation, reflections upon its formal conditions of possibility. This becomes clear as soon as we approach another level in *Remembrance,* the level of form—style, imagery, and syntax.

Machinations of Style

One summer day, the narrator is mounted on a horse in the Balbec countryside, riding along a quiet path in the woods near the sea. The beauty of the rugged landscape makes him think of two watercolors by Elstir, "Poet Meeting a Muse" and "Young Man Meeting a Centaur." The parallels are so striking that he feels as though he were traveling through a mythological landscape signed by Elstir. All of a sudden, the horse comes to a halt. The narrator catches sight of a supernatural creature:

> Suddenly, my horse reared; he had heard a strange sound; it was all I could do to hold him and remain in the saddle; then I raised my tear-filled eyes in the direction from which the sound seemed to come and saw, not two hundred feet above my head, against the sun, between two great wings of flashing metal which were bearing him aloft, a creature whose indistinct face appeared to me to resemble that of a man. I was as deeply moved as an ancient Greek on seeing for the first time a demi-god. I wept—for I had been ready to weep the moment I realised that the sound came from above my head (aeroplanes were still rare in those days), at the thought that what I was going to see for the first time was an aeroplane. (*REM* 2:1062/*RTP* 3:417)

For all its suddenness, Proust has carefully prepared the reader for the event. Just like Elstir's young man encounters a centaur and the poet runs into a muse, so Proust's hero is confronted with an airplane. The sight so overwhelms the narrator that he bursts into tears. To underscore the significance of the incident, Proust creates an almost archetypal view of the emergence of modernity; the rural stillness is played off against the noise of the airborne creature, and an old-fashioned means of transportation—

the horse—is contrasted to an ultramodern one—the airplane. But there is more to the passage. In juxtaposing the narrator's awe-inspiring experience with that of an ancient Greek, Proust aligns what is commonly identified as the beginnings of Western civilization with the narrator's present. The shuddering spectator on horseback thus faces the dawn of a new age, and history begins anew.

Terror, thrill, delight: Proust's airplane is sublime in Burke's sense, but there is a crucial difference—this is the technological sublime. As Rosalind Williams has shown, in the industrialized West the sublime branches out into the technological sublime, a notion whose meanings are endowed with a rich and nearly independent history.[49] Proust's airplane episode incidentally bristles with all of the exaltation that pervades Franz Kafka's "The Aeroplanes at Brescia" (1909). Written shortly after Blériot's flight across the English Channel, Kafka's piece relates an international aviation contest and how mighty wings glisten in the evening sun, including those of Blériot's famous monoplane. Characteristically enough, Kafka renders the day at the Italian airfield in the present tense, as though the breathtaking experience were beyond comparison.[50]

Proust's airplane episode, related in *Cities of the Plain*, bespeaks how the modernization process nibbles at the past. The present is split into past and future, old and new. Yet the appearance of the plane is not represented as doing violence to older mental processes. As the narrator underscores, the pilot's fearless flight across the sky signifies freedom (*REM* 3:100/*RTP* 3:612). In fact, the airman's maneuvering turns into a metaphor of sovereignty:

> I felt that there lay open before him—before me, had not habit made me a prisoner—all the routes in space, in life itself; he flew on, let himself glide for a few moments over the sea, then quickly making up his mind, seeming to yield to some attraction that was the reverse of gravity, as though returning to his native element, with a slight adjustment of his golden wings he headed straight up into the sky. (*REM* 2:1062/*RTP* 3:417)

The pilot comes across as a modern version of Icarus, the mythical figure who attempted to fly using wings made of feathers and wax. Icarus flew too close to the sun, the wax melted, and the young man plunged into the sea. Will Proust's pilot, too, go down? One thing is certain: the pilot is an emblem of the narrator's desire to leap into the future, the very artistic future that Proust suspends, and has to suspend, until the novel comes to an end.

In *Remembrance,* a striking difference makes itself felt between those passages that dwell on photography and cinema and those that deal with modern means of transportation. When the narrator renders trains, airplanes, and automobiles, he does so in an unmistakably enthusiastic spirit.[51] The difference is not quantitative, but qualitative; it points to two radically different thematic strands in the novel. The one strand is connected to the thematics of time, memory, and the past. This is the context in which Proust's narrator reflects upon photography and film. He uses them as defining counterpoints when working out his theory of involuntary memory, stressing over and over that they are deficient techniques for visual mimesis and mnemonic storage. In other words, photography and film belong to the economy in which Proust's official subject matter is situated: the thematics of writing, temporality, and involuntary memory.

As soon as memory falls outside the scope of the narrator's concerns, however, a quite different picture emerges. Photography and film cease to be problematic, and Proust might well make use of photographic means of representation, in particular when seeking to articulate acts of pure sensory perception. This, then, is the other major thematic strand in Proust's novel. It revolves around matters of perception and is concerned with the future, the narrator's writerly future.

Forget memory, Gilles Deleuze urges in *Proust and Signs* (1964). What makes Proust's tale coherent is not memory but the narrator's search, as in *recherche.* Approaching Proust's novel as the story of an apprenticeship of a man of letters, Deleuze suggests that it chronicles how the narrator acquires reading skills that enable him to move through and understand widely different worlds of signs.[52] Concerned with everything except the reanimation of the past, the narrator's search is thus directed at the future.

To this idea I would like to add another one. If the narrator makes his way through an opaque universe that demands to be read, it is not only because he is a novice trying to grasp the way of the world; it is also because the world itself is new. When a metal bird appears above the treetops in Balbec, the strange phenomenon in the sky—soon to be identified as an airplane—is initially represented as an enigma, as something illegible. For a few moments, the gap between signifier and signified lies open. Such instants are treasured in Proust, for when the material world emerges like a book written in a foreign language, the narrator turns into a reader; and at the same time, the world becomes writable once more. Indeed, a significant part of the narrator's apprenticeship in the art of reading signs takes place as he experiences the gradual breakthrough of modernity. As we shall see, technologies of speed are particularly instructive. And if an automobile makes the surrounding landscape undergo a metamorphosis, this is in keeping with a vital component of the aesthetic sensibility that *Remembrance* feeds on: that the artist should render what he sees and not what he knows.

Modernist Fictions of Speed

The higher the velocity, the more fatiguing travel becomes, Ruskin maintained. For Proust, the opposite was true. Speed quickened the eye. As Carter has demonstrated in *The Proustian Quest* (1992), Proust embraced the new culture of velocity with unmistakable enthusiasm, and his writings testify to this interest.[53] In fact, Proust declared himself a *fervent d'automobilisme,* a motoring fan.[54] Even when he traveled in the tracks of the British art critic whose work and sensibility he had admired for so long, he often chose to go by car, turning the ride into an aesthetic experience of the first rank. In 1907, when he motored in Normandy, automobiles were still something of a novelty. In 1900, as Pär Bergman has observed, there were only about 3,000 automobiles in France, a figure that had risen to 50,000 by 1909, and to 100,000 by 1913.[55]

Like numerous writers, painters, and photographers, Proust marveled at how speed helped transform the perception of time and space. In 1904, for example, the French photographer Robert Demachy captured a motorcar racing through a curve on a country road, and soon afterwards, the image was reproduced in Alfred Stieglitz's avant-garde art review, *Camera Work* (figure 8). Entitled *Vitesse,* the photograph implicitly addresses a problem that engaged numerous artists at the time: How is an image to convey an impression of speed and movement? That is to say, how is an image to represent temporal extension? Moreover, can the lived experience of locomotion be put into words?

That same year the Belgian writer Maurice Maeterlinck wrote a short story about a motoring trip in the countryside. Thanks to the car, the ecstatic narrator suggests, it has become possible for modern humans to absorb "in one day, as many sights, as much landscape and sky, as would formerly have been granted to us in a whole life-time."[56] Set in the neighborhood of Rouen, "In an Automobile" relates how the narrator goes for a drive by himself for the first time. He speeds past sleepy villages like a missile, cleaving yellow cornfields and red poppy meadows. Traveling at a speed of, say, thirty kilometers an hour, the vehicle turns the quiet country road into a spectacle:

8. Robert Demachy, *Vitesse*, 1904. Photogravure. Private collection.

The pace grows faster and faster, the delirious wheels cry aloud in their gladness. And at first the road comes moving towards me, like a bride waving palms, rhythmically keeping time to some joyous melody. But soon it grows frantic, springs forward, and throws itself madly upon me, rushing under the car like a furious torrent, whose foam lashes my face [. . .]. It is as though wings, as though myriad wings no eye can see, transparent wings of great supernatural birds that have their homes on invisible mountains swept by the eternal snow, have come to refresh my eyes and my brow with their overwhelming fragrance! (181–82/60–61)

Inert matter is here endowed with anthropomorphic features. All that seemed solid jumps into motion. In this way, Maeterlinck attempts to convey a sense of the vertiginous speed with which the motorcar travels through space. For despite all its impressive horsepower, it is not the car but nature that acquires agency in Maeterlinck's scenario. Nature, not the car, emerges as a reckless force apparently beyond masculine control.

Revolving around the thrill of the new, Maeterlinck's short story emphasizes the miracle of speed and the concomitant transformations of space. As the vehicle plunges into a valley, the velocity animates the poor old trees lining the road:

The trees, that for so many slow-moving years have serenely dwelt on its borders, shrink back in dread of disaster. They seem to be hastening one to the other, to approach their green heads, and in startled groups to debate how to bar the way of the strange apparition. But as this rushes onward, they take panic, and scatter and fly, each one quickly seeking its own habitual place; and as I pass they bend tumultuously forward, and their myriad leaves, quick to the mad joy of the force that is chanting its hymn, murmur in my ears the voluble psalm of Space, acclaiming and greeting the enemy that hitherto has always been conquered but now at last triumphs: Speed. (182–83/61–62)

For all its mundanity, the event is of world-historical significance. Thanks to speed, Maeterlinck asserts, man has triumphed over time and space (183/62). The dust of history whirls up behind the motorcar, and nothing will ever be the same.

In 1906, another Belgian writer, Eugène Demolder, published a literary description of an exhilarating motoring trip through Spain, *L'Espagne en auto*.[57] The automobile races through the picturesque landscape as though for the first time, producing one visual thrill after another. Here, too, speed is the protagonist, evoked by an army of exclamation marks and a breathlessly telegraphic prose.

The French writer Octave Mirbeau also conceived a story glorifying the horsepower of the motorcar. Mirbeau's 1907 novel *La 628–E 8*, so named after his license plate, relates a motoring trip through France. In the opening pages, the narrator turns speed into an emblem of human consciousness:

His brain is a racetrack around which jumbled thoughts and sensations roar past at 60 miles an hour, always at full throttle. Speed governs his life: he drives like the wind, thinks like the wind, makes love like the wind, lives a whirlwind existence. Life comes hurtling at him and buffeting him from every direction, as in a mad cavalry charge, only to melt flickeringly away like a film [*et disparaît cinématographiquement*] or like the trees, hedges and walls that line the road. Everything around him, and inside him, dances, leaps, and gallops, in inverse proportion to his own movement; not always a pleasant sensation, but powerful, delirious and intoxicating, like vertigo or fever.[58]

Diagnosing life in the modern age, Mirbeau suggests that two features characterize the world of the living: an overwhelming experience of speed and an equally overwhelming mass of visual stimuli. Trees, walls, silhouettes—everything is flickering before the eyes of modern humans. Described as a succession of moving images, the world emerges as a film.

Maeterlinck, Demolder, and Mirbeau explore the syntax of velocity, seeking to convey an experience to which the medium of the printed word is stubbornly resistant: the bodily experience of speed. They thus contemplate a representational problem: how the formal level of description may call forth what is suggested on the level of enunciation. And in response to this linguistic challenge, they settle on a rhetoric of inversion, substituting the immobile for the mobile, and vice versa. Mirbeau even writes that the movement of human beings stands in inverse proportion to that of the environment.

It is no coincidence that Mirbeau uses cinema as a metaphor of how the world incessantly races forward and disappears behind the human subject, for it is film that offers a model for how to represent the lived experience of velocity. Early cinema specifically experimented with ways of creating panoramas that simulated speed and movement. From a historical point of view, Tom Gunning argues, "camera movement began as a display of the camera's ability to mobilize and explore space."[59] Up until 1908, when so-called narrative cinema was more fully developed, a vast number of filmmakers explored speed and movement by mounting a camera on vehicles such as subway cars, streetcars, trains, aerial balloons, and gondolas. Predicated on the inversion of mobility and immobility, this widespread shooting technique produced an optical illusion of movement by suggesting that the environment rushed towards the camera and, by implication, the spectator.

As early as 1895, shortly after the first screenings of the Lumière program, the writer H. G. Wells and his collaborator Robert Paul had plans to construct a cinematic device for simulating movement through time and space, a visual attraction on the order of Wells's 1895 science-fiction novel *The Time Machine*.[60] The project failed, but the first decade of the twentieth century saw the emergence of widely popular fairground attractions that similarly produced an illusion of speed and movement. At the Paris Exposition in 1900, the *cinéorama* was introduced, simulating the experience of traveling in an aerial balloon hovering over Europe and Africa.[61] The Exposition also included the *maréorama,* which simulated sea travel—hence the name (figure 9). Going from Villefranche to Constantinople via Naples and Venice, the trip came complete with sirens going off and black smoke rising to the sky (figure 10). A guaranteed success, commented *Le Figaro's* enthusiastic correspondent on 22 July 1900.[62]

9. Advertisement for the Maréorama at the Exposition Universelle de 1900.
Getty Research Institute, Research Library, Los Angeles.

A few years later, an American fairground attraction called Hale's
Tours was invented. A worldwide commercial success, especially in the
years 1905–7, it offered a synaesthetic experience of railway travel around
the world. Audiences were seated in a railroad car with an open front,
which served as a cinematic frame for motion pictures that had been shot
from the cowcatcher of a train. The swaying of the railway car, blowing
winds, starts and stops, and the characteristic clickety-clack sound en-
hanced the experience of virtual movement.

The train and the cinema are two distinct sites of spectatorship, to be
sure, but as Lynne Kirby has argued in her work on early cinema, they
share a fundamental affinity.[63] In fact, Kirby suggests that the railroad
paved the way for the specifically cinematic construction of the spectator.
Add to this historical scenario the advent of the automobile, yet another
visual framing device on wheels, and we begin to appreciate the optical
thrill that Proust's narrator associates with technologies of speed.

In 1908, Leo Tolstoy suggested that "cinema has divined the mystery of
motion."[64] In 1909, F. T. Marinetti declared that "the world's magnificence
has been enriched by a new beauty; the beauty of speed."[65] The first futurist manifesto was published on 20 February 1909 on the front page of *Le
Figaro*, proclaiming, among many other things, the superior beauty of the
motorcar. Fifteen months earlier, on 19 November 1907, the same newspa-

10. Technical construction of the Maréorama. *La Nature*, no. 1414 (30 June 1900). Princeton University Library.

per had featured a piece called "Impressions de route en automobile," also on the front page, and the author of the article was Marcel Proust.

The Motoring Fan and the Simulacrum of Velocity

A stylistic exercise in modernist techniques of visual mimesis, "Impressions de route en automobile" addresses a representational problem: how to render the lived experience of speed and movement. Written in the early fall of 1907, the piece is based on the motoring trips Proust made that same summer with Alfred Agostinelli, the driver he had hired for the purpose. Together they visited places in Normandy that Ruskin had written about, such as Bayeux, Conches, Lisieux, and Caen. Comfortably reclining in the passenger seat, Proust was able to concentrate on the vistas passing by; the most thrilling ones made their way into "Impressions de route en automobile." In fact, Proust was so pleased with the centerpiece of the article—the description of the church steeples rising above Caen—that he was to recycle it in *Remembrance*, in the pages known as the Martinville episode.[66] Widely thought of as one of the most important and most enigmatic passages in the novel, the episode relates how the young narrator, sitting next to the coachman on a horse-driven carriage, comes to witness how the steeples of Martinville dance on the horizon. Significantly enough, the visual experience prompts the narrator's very first piece of writing.

As Jean-Yves Tadié has suggested, the *Figaro* article was an "embryo of the future work."[67] The article merits our attention, and not just because it casts an interesting light on the genesis of Proust's novel. Read closely, it complicates the received view of the French author's relationship to Ruskin. Although Proust's *période ruskinienne* is commonly believed to have ended in 1906, the motoring article shows that post-Ruskinian Proust was at once more Ruskinian and more modernist than has been thought. As we shall see, Proust paid tribute to his old mentor's aesthetic program in rather unexpected ways, by turning Ruskin's antimodern stance on its head. Ultimately, the 1907 article provides us with a richer understanding of the aesthetics of perception that informs *Remembrance of Things Past*.

The piece relates how the narrator and his driver travel along a road in Normandy, going from a place near Caen to a small town between Lisieux and Louviers. The purpose of their trip is to pay a visit to the narrator's parents. The purpose of the article, however, is to depict how the motorcar catapults through space and how speed transforms the surrounding landscape into a phantasmagoria. These descriptions are then flanked by digressions on architecture, religious art, and the history of travel. Both Ruskin and Turner are mentioned, as are seventeenth-century Dutch painters such as Philips Wouwermans and Adriaen van de Velde. Proust thus makes sure to anchor his speed-infused landscape in the esteemed traditions of art history.

No wonder, then, that Proust's narrator so often resorts to a visual vocabulary. At the outset, he sets the stage for the optical adventures that will follow by pointing out that the windshields convert the vista into an aesthetic object: "To my right, to my left, in front of me, the car windows—which I kept closed—put behind glass, so to speak, the beautiful September day which, even in the open air, one could only see as through a kind of transparence" (*CSB* 63). Darting across the plain of Caen at sunset, the motorists experience a visual event whose delightful nature is a function of the speed. The windshields delimit the view of the landscape, transforming it into an object of visual pleasure—a mobile panorama. As Wolfram Nitsch suggests, the motorcar has become a "vehicle of perception," thus paving the way for new forms of aesthetic gratification, much in the spirit of Elstir.[68]

Meanwhile, it is as though the motorcar is at a standstill while the environment rushes toward it. No sooner has the vehicle gathered momentum than the motorists come to witness how ancient buildings spring to life: "old lopsided houses ran nimbly toward us, offering us fresh roses or proudly showing us the young hollyhock they had grown and which already had outgrown them" (*CSB* 63). The inanimate becomes animate; the immobile becomes mobile. Similarly, when the motorcar arrives at the foot of the cathedral in Caen, Proust writes not that the automobile speeds toward it; instead, it is the steeples that throw themselves at the vehicle, and with such violence that the driver almost collides with the porch. Like Maeterlinck and Mirbeau, Proust reverses mobility and immobility in his search of a rhetoric of speed capable of evoking the velocity with which the motorcar traverses the plain.[69] At the same time, Proust insists on what the motorist perceives and not on what he knows. This stylistic technique animates the central panel of the article, the artful representation of the steeples rising above Caen:

The minutes passed, we were travelling fast, and yet the three steeples were always ahead of us, like birds perched upon the plain, motionless and conspicuous in the sunlight. Then the horizon was torn open like a fog unveiling completely and in detail a form invisible an instant before, and the towers of the Trinity appeared, or rather what appeared was one single tower perfectly hiding the other. But it drew aside, the other stepped forward and both of them lined up. Finally, a dilatory steeple [. . .] had come to join them, springing into position in front of them with a bold leap. Now, between the propagating steeples below which one saw the light which at this distance seemed to smile, the town, following their momentum from below without being able to reach their heights, developed steadily by vertical rises the complicated but candid fugue of its rooftops.[70] (*CSB* 64)

The motorized spectator surely knows what the sun-drenched phenomena on the horizon represent, but he presents them as synecdoches, as part of a not-yet-revealed whole—as steeples, not churches. The steeples are a sensuous sign, an enigmatic piece of writing whose meaning awaits decipherment. In this sense, too, the narrator insists on what his eyes perceive and not on what he knows, all in an effort to render the lived experience of speed and the delicious perception of the landscape through which the car races:

We had long since left Caen, and [. . .] the two steeples of Saint-Etienne and that of Saint-Pierre waved once again their sun-bathed pinnacles in token of farewell. Sometimes one would withdraw, so that the other two might watch us for a moment still; soon I saw only two. Then they veered one last time like two golden pivots, and vanished from my sight. Many times afterwards, when travelling across the plain of Caen at sunset, I would see them again, sometimes at a great distance, and seeming no more now than two flowers painted upon the sky above the low line of the fields; sometimes from a little closer, the steeple of Saint-Pierre having already caught up with them, they seemed like three maidens in a legend, abandoned in a solitary place over which night had begun to fall; and as I drew away from them, I could see them timidly seeking their way, and after some awkward attempts and stumbling movements of their noble silhouettes, drawing close to one another, gliding one behind another, forming now against the still rosy sky no more than a single dusky shape, pleasant and resigned, and so vanishing in the night. (*CSB* 65)

"The psychology of motoring has found its true writer in Proust," Ernst Robert Curtius maintains in his classic study of *Remembrance*. Reflecting on the close link between the narrator's epiphanic moments of inspiration and his speedy movement through space, in "Impressions de route en automobile" as well as in *Remembrance*, Curtius aptly remarks that one cannot imagine Proust on foot.[71] Indeed, in Proust speed bursts open a new universe. As the immobile is set in motion, the artistic eye takes possession of the newly transformed visual world.

The Steeples of Martinville: Birth of a Writer

In the Martinville episode in *Swann's Way*, Proust makes use of the motoring article he wrote for *Le Figaro*.[72] In these inspired pages, however, he drops the narrative about the automobile ride, along with the art-historical references and the digressions on travel. All that remains is the central panel: the mighty vision of the mobile steeples. The picturesque panorama is now placed within another story, one that revolves around the young narrator's fear that he might never become a writer.

One evening, the young hero travels along the Guermantes way in Dr Percepied's carriage. Sitting next to the coachman, he comes to witness how the glistening steeples of Martinville turn into anthropomorphic characters before his eyes. The sight arouses a peculiar pleasure. He decides to act on the irksome impulse, borrows a pencil and a piece of paper from the doctor, and begins to write. It is his first piece of writing. The emergence of writing, then, is intimately linked to technologies of velocity and the new spaces of representation they burst open.

Later, looking back at the event, the older narrator details the optical scenario, how the road bends, how the twin steeples appear to change position, now here, now there, and how, suddenly, they are joined by a third sunlit steeple. Remarkably, he then proceeds to duplicate the description of the vertiginous ride and the enchanted steeples. Halting the diegetic flow, he presents the episode to the reader again—in the form of the fragment drafted on the coach box. It is the one and only time that the narrator quotes himself, and the repetition, of course, underscores the singular importance of the sequence. As Keith Cohen has observed, "Marcel's experience of seeing the [. . .] steeples from a moving coach makes an indelible mark on the text, since it gives rise to the first piece of writing *within* the text attributed to Marcel himself."[73]

Apart from having been adapted to the Combray topography, and apart from a few cuts and minor modifications, the fragment written on the coach box is a word-for-word replica of the centerpiece of the article on the visual pleasures of motoring written in 1907. As Proust himself was later to point out, and as Bernard Guyon has shown in detail, the passages are virtually identical.[74] The only significant difference is that Proust has

The Education of the Senses **133**

substituted an old-fashioned carriage for the original automobile. There is an obvious reason. The Martinville episode takes place in the early 1880s, well before the advent of the automobile. Presumably the narrator has traveled in a carriage many times, so how can it be that he experiences such perceptual freshness? As Guyon suggests, this is why Proust underscores the speedy gallop and Dr Percepied's exceptional haste.[75] Hence the dewiness of the narrator's visual experience.

The Martinville episode rounds off *Combray*, the overture to *Remembrance*. But there is more to the episode, for it is to reappear as a leitmotiv within Proust's narrative as a whole, as part of the overarching theme that revolves around the narrator's anxiety that he will fail to become a writer. If one pulls the thread that Proust lays out in *Swann's Way* and that runs through the novel up until the penultimate volume, one will discover some of the stitches that hold the unwieldy tale together. In *The Guermantes Way*, the narrator mentions in passing that he has found the piece about the Martinville steeples that he wrote on Dr Percepied's coach box and that he has submitted a slightly edited version to *Le Figaro* (*REM* 2:412/*RTP* 2:691–92). It seems to have been in vain, however, for the article fails to appear. In *The Captive*, the narrator relates twice how he opens *Le Figaro* in the hope of discovering his article in print (*REM* 3:4–5, 114/*RTP* 3:523, 626).

It is only in *The Fugitive*, the next-to-last volume, that the plot about the unprinted article reaches its climax and the piece finally appears in the pages of the Parisian paper. The narrator swells with pride. He is so taken with the dignity of the event that he begins to sing the praises of the word, the mass-produced word:

> I considered the spiritual bread of life that a newspaper is, still warm and damp from the press in the murky air of the morning in which it is distributed, at daybreak, to the housemaids who bring it to their masters with their morning coffee, a miraculous, self-multiplying bread which is at the same time one and ten thousand, which remains the same for each person while penetrating innumerably into every house at once. (*REM* 3:579/*RTP* 4:148)

The daily paper is to the soul what bread is to the stomach. In Proust's hands, the newspaper—this profane piece of printed matter that, at best, enjoys a twenty-four-hour lifespan—acquires almost religious dimensions. The passage alludes to the episode in the New Testament where Jesus miraculously produces bread for the hungry masses gathered to listen to his sermon. Rarely has a daily paper been described in loftier terms.

And this in a work of literature that so emphatically celebrates the *book*—and certainly not as a mass product but as a spiritual phenomenon.[76]

The leitmotiv weaves its way through a few more pages. Monsieur de Guermantes congratulates the narrator on a splendid piece of writing and on having found himself an activity worthy of esteem, but he laments the somewhat precious and clichéd style that, to his mind, characterizes large parts of the text. Tough criticism, to be sure, but what is essential is that the narrator has won recognition as a writer. The thematic thread thus binds together a course of events that begins with the Martinville episode, when the narrator produces his first piece of writing, and that ends when *Le Figaro* finally runs the article and he makes his long-desired debut. Only one thing remains: for the narrator to write a significant work of literature.

When *Remembrance* was in its beginning stages, Proust considered letting the novel open with the episode where the narrator sees his first publication in print. In these early sketches, the narrator turns around in his bed, just as in the final version. The difference is that he is anxiously contemplating the fate of the article he has long since submitted to *Le Figaro*. The next day his proud mother, trying in vain to appear casual and indifferent, enters his bedroom with a copy of *Le Figaro* featuring his article on the front page. Although Proust was to move the *Figaro* episode from the opening pages of *Swann's Way* to *The Fugitive*, the fact that he originally thought of linking the bedtime episode that opens the novel to the *Figaro* episode affirms the latter's importance (*RTP* 1:633–36, 639).

The Aesthetics of the Windshield

At the time of "Impressions de route en automobile," as I have suggested, we find Proust traveling again in Ruskin's footsteps, more than a year after his Ruskinian period is typically believed to have come to a close.[77] To prepare himself for these excursions, Proust asked his friend and ex-partner Reynaldo Hahn to send him a number of Ruskin volumes.[78] It should not come as a surprise, then, that the piece of prose fiction that resulted from these trips treats a number of typically Ruskinian themes: landscape, travel, architecture, painting, seeing; in addition, Ruskin is mentioned twice and quoted once. In fact, as Autret has maintained, "Impressions" was Proust's last Ruskinian article.[79]

It is no doubt true that Ruskin's spirit pervades "Impressions," at least as long as the car is kept out of the picture. But the car, of course, is the motor of the scenario. It is both part of the spectacle and its very vehicle.

Take away the car and the speed-infused panorama loses its charge, for it is the not-yet-domesticated experience of velocity that, along with the window frames, give rise to the visual marvels that Proust so painstakingly describes. Thanks to the automobile, a new space of representation has made itself available to aesthetic exploration.

The British painter and Royal Academician Hubert von Herkomer once explained the appeal of the automobile: "The pleasure [of motoring] is seeing Nature as I could in no other way see it; my car having 'tops,' I get Nature framed—and one picture after another delights my artistic eye," he wrote in 1905.[80] As Herkomer gazed through the windshield, the view turned into a series of tableaux complete with frames. The automobile had become a viewing instrument on the order of, say, the Claude glass.

It is hard to think of a more congenial representation of this idea than Henri Matisse's *Le Pare-brise: Sur la route de Villacoublay* (The Windshield). Conceived during a motoring trip in southern France in 1916, Matisse's painting takes as its subject a windshield, including the view of the road and the surrounding landscape as seen through the windows (figure 11). A study in perspective, the painting foregrounds the way in which the view is framed, exploring how the three windows delimit the spectator's field of vision and divide the seen into separate yet related visual spaces. These spaces, in their turn, transform themselves into a pictorial suite, into three distinct images—it is as though they were only waiting to be lifted into the artist's sketchbook. And this is indeed what is taking place in Matisse's painting. In the right-hand corner, a sketchbook is propped up against the steering wheel. The frame of the sketch under way corresponds to the window frame in the front which, in its turn, corresponds to the frame of the canvas itself. Such a complicated play of frames underscores that the painting does more than just depict the road from Villacoublay; it also records its own coming into being. Indeed, *Le Pare-brise* is a meta-painting, ultimately turning the automobile into a metaphor for the painter's eye.

"When one crosses a landscape by automobile or express train, it becomes fragmented; it loses in descriptive value but gains in synthetic value," Fernand Léger wrote in 1914. Contemplating the affinity between modern technologies of speed and new ways of seeing, he suggests that the "view through the door of the railroad car or the automobile windshield, in combination with the speed, has altered the habitual look of things."[81] Proust's mode of describing the automobile ride, the excitement of speed, and the dazzling views through the windshield is a brilliant example of precisely the kind of modernist aesthetic that Léger promotes in

11. Henri Matisse, *Le Pare-brise: Sur la route de Villacoublay*, 1916/1917. Cleveland Museum of Art, bequest of Lucia McCurdy McBride in memory of John Harris McBride II, 1972. © 2001 Succession H. Matisse, Paris / Artists Rights Society (ARS), New York.

his essay, written half a dozen years after Proust's motoring piece was first published.

We have seen how Proust, in embracing the modern and making it a vital part of his visual aesthetics, effectively negates Ruskin's antimodern spirit.[82] In a 1909 pastiche, Proust continues his gentle mockery of Ruskin's programmatic critique of all things mechanical. He imagines that the British critic has penned a description of Paris as seen from an airplane, a creation that, as Proust writes in a humorous aside, "is rightly known as one of the master's most refined pieces."[83] Parodying Ruskin's style, Proust describes sight, sites, and atmospheres using the expansive idiom so characteristic of Ruskin's works. These poetic impressions, however, are not generated by the art critic's pretechnological eye. Indeed, this is not a Ruskin who takes slow walks through the countryside but one who darts across the skies in a vision machine, reveling in the joy of looking down on the domes of Sacré-Cœur. And when the mechanical bird finally descends, the creative gaze of the airborne spectator perceives not the Eiffel Tower but the spear of Odin. In Proust's pastiche, air travel thus promotes precisely the kind of fresh sensory experience that Ruskin honored.

Distant echoes of the airplane motif can be heard in *The Fugitive,* in the episode where the narrator views Giotto's frescoes in the Arena chapel in Padua. Gazing in amazement at the famous cycle on which Ruskin dwells in *Giotto and His Works in Padua,* the narrator is struck not by the artist's treatment of the Annunciation, the Nativity, the Crucifixion, and the other biblical events that make up the glorious life of Christ. Instead, what catches his eye are Giotto's angels, those speedy creatures criss-crossing the heavenly spheres with all the compelling athleticism other-wise associated with early twentieth-century aviation shows:

> Constantly flitting about above the saints whenever the latter walk abroad, these little beings, since they are real creatures with a genuine power of flight, can be seen soaring upwards, describing curves, "loop-ing the loop" [*décrivant des courbes, mettant la plus grande aisance à exécuter des loopings*], diving earthwards head first, with the aid of wings which enable them to support themselves in positions that defy the laws of gravity, and are far more reminiscent of an extinct species of bird, or of young pupils of Garros practising gliding, than of the an-gels of the Renaissance and later periods whose wings have become no more than emblems and whose deportment is generally the same as that of heavenly beings who are not winged. (*REM* 3:663/*RTP* 4:227)

Without further ado, Giotto's angels are here turned into airplanes en-gaged in a sophisticated air show à la Roland Garros. According to Paul Robert's dictionary, this is the first time the word *looping,* in the sense of "looping the loop," appears in French.[84] Ruskin surely would have turned over in his grave at the thought of Giotto's heavenly hosts being likened to airplanes.[85] Yet, in insisting on the aerotechnical gusto with which they race across the sky, Proust pays tribute to Giotto, to all the muscular vigor that characterizes his treatment of the angels. As opposed to their ethereal Renaissance colleagues, Giotto's able-bodied creatures have been fur-nished with real wings and can actually fly. A greeting from one artist to another, the episode celebrates the flesh-and-blood naturalism that has ensured Giotto such a prominent place in the annals of art history. Clearly, Proust more than cancels Ruskin's antimodern stance; he also works out a vital element of his own aesthetics.

Machines of Vision

The *Figaro* article provided Proust with a rhetoric of movement that he was to make use of numerous times in *Remembrance.* In *The Fugitive,* for example, when the narrator and his mother are on their way back to Paris from Venice, the train gathers speed so quickly that the northern Italian landscape comes to life and turns into a mobile panorama: "The train started and we saw Padua and Verona come to meet us, to speed us on our way, almost on to the platforms of their stations, and, when we had drawn away from them, return—they who were not travelling and were about to resume their normal life—one to its plain, the other to its hill" (*REM* 3:670/*RTP* 4:234). From Venice to Verona via Padua: a distance of one hundred twenty kilometers is covered in a single sentence, and the velocity is illustrated by a billowing movement that is attributed not so much to the train as to the age-old cities along the way.

An even more striking example of Proust's rhetoric of inversion is to be found in *Cities of the Plain,* in an episode partly inspired by the *Figaro* article. The narrator and Albertine have decided to go on a motoring trip. The passage suggests that it happens for the first time, for much to their surprise, the driver tells them that it will take a mere twenty minutes to travel to Saint-Jean, and from Quetteholme to la Raspelière only thirty-five minutes. "We realised this as soon as the vehicle, starting off, covered in one bound twenty paces of an excellent horse" (*REM* 2:1029/*RTP* 3:385). For yet a little while, the velocity of the horse is a given point of reference; and it is this older mental space that makes possible the perceptual adventures that will soon unfold. But first the narrator supplies a philosophical reflection upon the vocabulary of space-time: "Distances are only the relation of space to time and vary with it. We express the difficulty that we have in getting to a place in a system of miles or kilometres which becomes false as soon as that difficulty decreases. *Art is modified by it also,* since a village which seemed to be in a different world from some other village becomes its neighbour in a landscape whose dimensions are altered" (*REM* 2:1029/*RTP* 3:385; emphasis added).

The faster humans travel through space, the smaller the world becomes, the narrator suggests, for thanks to the motorcar, one may visit both Saint-Jean and La Raspelière in one afternoon, two places that used to be "as hermetically confined in the cells of distinct days as long ago were Méséglise and Guermantes." No wonder, then, that Padua and Verona appear side by side in one and the same sentence; the speeding train has shoved them together.

Not only is the world becoming smaller; the passage states that art, too, is affected by these changes. Indeed, embedded in this proto-materialist reflection upon the historicity of perceptual habits is a metacommentary, a discourse within the discourse that ultimately casts a historical light on

Proust's own aesthetic endeavor. Such metacommentaries form a wide-ranging subterranean system in *Remembrance*. The narrator has already observed, for example, that the railway has taught modern humans to take account of minutes, that the senses grow stronger the more one makes use of them, that the advent of photography has elicited Elstir's aesthetics, and that the telephone affects habitual ways of perceiving voices and faces. The narrator knows that his artistic enterprise partakes of this history. To be sure, he emerges as an impassioned idealist, pursuing as he does those fragments of time in the pure state and celebrating the work of art as a transcendental entity. But the narrator is also inhabited by a creature with a highly developed historical-materialist sensibility. The tension between these two impulses—idealism and materialism—animates Proust's tale from beginning to end.[86]

The suggestion that modern means of transport such as the motorcar affect habits of perception is then immediately echoed in the description that follows when the narrator takes us back to the automobile ride to La Raspelière: "Coming to the foot of the cliff road, the car climbed effort-lessly, with a continuous sound like that of a knife being ground, while the sea, falling away, widened beneath us. The old rustic houses of Montsur-vent came rushing towards us, clasping to their bosoms vine or rose-bush; the firs of la Raspelière [. . .] ran in every direction to escape from us" (*REM* 2:1029–30/*RTP* 3:386).

The proper content of the passage is the excitement of speed, the throbbing thrill of the new and the estrangement of the familiar. And if the scenario appears before the reader as though newly caught, it is because Proust insists on the Ruskinian belief in the primacy of the perceived. He does so particularly on the syntactic level. Not only is the motorcar turned into a subject; the vehicle is soon accompanied by other inanimate objects that also spring to life and acquire syntactical agency: the sea *pulls back* behind the automobile, the houses *rush* toward the motorists, the firs *leap* to the side like anxious pedestrians. Here as in "Impressions," Proust deploys the figure of inversion, and the qualities normally attributed to automobiles—force, speed, mobility—attach themselves to the immobile environment.[87] What is more, the passage shows Proust at his most modernist, for at the center stands not a human subject but a motorcar, celebrated by the futurists as the true incarnation of beauty, second only to the airplane. Proust makes use of a similar reversal technique when describing a gondola trip in Venice. At first the emphasis is placed on the craft and how it moves through the neighbor-hoods, and then the city itself comes to life (*REM* 3:641/*RTP* 4:206). Like the inspired steeples rising above Caen, the Venetian canals and façades turn into a mobile panorama. Habit has not yet domesticated the immediacy of sensory experience.

Yet although such passages show Proust at his most modernist, perhaps at his most innovative, his stylistic techniques are not without precedent. The Proustian mode of rendering speed and movement is closely related to representational techniques inherent in early cinematography and turn-of-the-century fairground attractions. Cinema enabled the perception of speed as a matter of inversion in the first place, offering at the same time its mode of representation.[88] That Proust developed what may be conceived of as literary counterparts of cinematographic techniques, especially montage and close-up techniques, was noted by several early reviewers.[89] Yet the resonance between Proust's style and cinema has been largely unexplored.[90] What has gone unnoticed is the tension between the narrator's critical reflections upon cinematography and the appropriation of uniquely cinematographic modes of signification. What has also been neglected is the complex relationship between the ideal of unmediated vision articulated by *Remembrance* and the use of photographic techniques to represent such visual experience.

Two final examples, both drawn from *The Guermantes Way,* will advance my argument. The narrator is about to kiss Albertine for the first time. Detailing how the narrator's mouth travels toward the loved one's face, the description requires some five pages. Just at the moment when those shivering lips are about to reach their destination, the action comes to a halt: "At first, as my mouth began gradually to approach the cheeks which my eyes had recommended it to kiss, my eyes, in changing position, saw a different pair of cheeks; the neck, observed at closer range and as though through a magnifying-glass, showed in its coarser grain a robustness which modified the character of the face" (*REM* 2:378/*RTP* 2:660).

The choice of metaphor—the magnifying glass—is no coincidence. Thousands of words have been used to describe the narrator's desire to be close to Albertine, and yet the passage speaks of alienation, of the alienating power of the gaze. This peculiar tension between proximity and distance is then amplified when the passage unexpectedly breaks off into a digression on . . . photography. Right in the middle of the action, the narrator goes off the tracks and decides to explore the remarkable ability of photography to produce new perspectives on the spatial location of objects:

Apart from the most recent applications of photography—which huddle at the foot of a cathedral all the houses which so often, from close to, appeared to us to reach almost to the height of the towers, drill and deploy like a regiment, in file, in extended order, in serried masses, the same monuments, bring together the two columns on the Piazzetta which a moment ago were so far apart, thrust away the adjoining dome of the Salute, and in a pale and toneless background manage to include a whole immense horizon within the span of a bridge, in the embrasure of a window, among the leaves of a tree that stands in the foreground and is portrayed in a more vigorous tone, frame a single church successively in the arcades of all the others—I can think of nothing that can to so great a degree as a kiss evoke out of what we believed to be a thing with one definite aspect, the hundred other things which it may equally well be, since each is related to a no less legitimate perspective. (*REM* 2:378/*RTP* 2:660)

In the middle of the kiss, the narrator offers a miniature essay on photography and its impact upon habitual modes of perception. Using Venice as an example, he suggests that photography elicits unexpected perspectives on familiar sights. The narrator does not state it, but there is a close affinity between technologies of speed and photography. Just as the motorcar shrinks the distance between remote villages, and just as the train turns Padua and Verona into neighbors, so photography makes buildings and monuments slide apart or squeeze together, step into the foreground or turn into a mere backdrop, all depending on the point of view. Photography estranges the familiar, and for this reason, the narrator suggests, the kiss is akin to photography. And so Albertine, only seconds away, is promptly turned into a spectacle, her face appearing as a series of moving façades:

> In short, just as at Balbec Albertine had often appeared different to me, so now—as if, prodigiously accelerating the speed of the changes of perspective and changes of colouring which a person presents to us in the course of our various encounters, I had sought to contain them all in the space of a few seconds so as to reproduce experimentally the phenomenon which diversifies the individuality of a fellow-creature, and to draw out one from another, like a nest of boxes, all the possibilities that it contains—so now, during this brief journey of my lips towards her cheek, it was ten Albertines that I saw; this one girl being like a many-headed goddess, the head I had seen last, when I tried to approach it, gave way to another. (*REM* 2:378–79/*RTP* 2:660)

Up close, Albertine's physiognomy is both decomposed and multiplied. What is implied here is that the narrator's gaze relates to the loved one's face in much the same way as the camera relates to a townscape. In other words, a photographic vocabulary serves to represent the narrator's seemingly unmediated visual experience of Albertine's appearance.

It has been suggested that such passages testify to Proust's essential modernness, as they demonstrate that he invented stylistic counterparts of cubist as well as futurist painterly techniques.[91] Duchamp's *Nude Descending a Staircase* is frequently cited as a point of reference. Some critics have even pointed out the possible influence of Marey's decomposition of bodily movement on Proust's stylized images, arguing that Proust created "literary chronophotographs."[92] This is no doubt true. Yet the moment one concludes that Proust cultivated a "chronophotographic" style, one has laid bare a paradox at the heart of his artistic enterprise—a paradox that merits serious consideration, as it complicates the commonly accepted view of Proust's aesthetics of perception. On the one hand, Proust's novel subscribes to the idea of unmediated visual experience. On the other hand, when seeking to articulate such experiences, he often stages them by means of photographically mediated modes of representation.

One final example: during a walk with Saint-Loup, the narrator gets so distracted by precious memories of his former girlfriend Gilberte that he lags behind and loses sight of Saint-Loup. All of a sudden, he becomes aware of a strange incident further down the sidewalk:

> I saw that a somewhat shabbily attired gentleman appeared to be talking to [Saint-Loup] confidentially. I concluded that this was a personal friend of Robert; meanwhile they seemed to be drawing even closer to one another; suddenly, as an astral phenomenon flashes through the sky, I saw a number of ovoid bodies assume with a giddy swiftness all the positions necessary for them to compose a flickering constellation in front of Saint-Loup. Flung out like stones from a catapult, they seemed to me to be at the very least seven in number. They were merely, however, Saint-Loup's two fists, multiplied by the speed with which they were changing place in this—to all appearance ideal and decorative—arrangement. But this elaborate display was nothing more than a pummelling which Saint-Loup was administering, the aggressive rather than aesthetic character of which was first revealed to me by the aspect of the shabbily dressed gentleman who appeared to be losing at once his self-possession, his lower jaw and a quantity of blood. (*REM* 2:186/*RTP* 2:480)

Brimming with perceptual excitement, the humorous passage opens with confusion. Soon enough the hermeneutic horizon begins to take shape, and the narrator—along with the reader—realizes what he has been witnessing all along: a scuffle! The course of events takes only a few seconds, yet it is as though the narrator has put them under a magnifying glass. The proper content of the passage is neither Saint-Loup nor the fist fight; it is, rather, the narrator's perceptual act as such—how he tries out hypothesis after hypothesis in an attempt to figure out the true nature of the event taking place before his eyes.[93] The narrator carefully reports what he sees, and with scientific precision: at least seven egg-shaped bodies, "flung out like stones from a catapult." These then turn out to have been a phantasmagoria, and Saint-Loup's angry fists emerge into view. The scenario thus moves from the narrator's unmediated visual impressions to secure positive knowledge.

Or so it would seem. On closer inspection, the description is less inductive than one may think. The narrator's seemingly spontaneous visual experience is refracted through a perceptual grid. In seeking to capture the nature of the optical illusion and represent its phenomenology to the reader, he draws on a readily available visual system of signification: chronophotography. Indeed, the description of the corporeal movements that make up the fistfight is enabled by the chronophotographic representation of bodies in motion.

Marey and Muybridge, we recall, had rendered visually what the naked eye had never been capable of perceiving; for example, all four hoofs of a horse levitating in the air at the same time. Challenging the epistemic powers of the eye, chronophotography broke down bodily movements into an arbitrary number of discontinuous positions, to so many frozen moments of spatialized time. The futurists, among others, adopted the chronophotographic language. In a 1910 manifesto, they explained the nature of futurist painting: "A profile is never motionless before our eyes, but it constantly appears and disappears. On account of the persistency of an image upon the retina, moving objects constantly multiply themselves; their form changes like rapid vibrations, in their mad career. Thus a running horse has not four legs, but twenty, and their movements are triangular."[94]

Proust's narrator similarly seizes on the new optical vocabulary articulated by Marey and Muybridge, and makes it a vital part of his visual aesthetics. Saint-Loup has not two fists but seven; Albertine's face is not one but ten. The narrator says: this is what it really looked like, this is how those visual impressions landed on my retina, this is how my sense of sight was laboring before reason arrived on the scene. And in order to reconstruct the immediacy of this emphatically individual and body-based perceptual act, the narrator turns to the seemingly objective language of chronophotography. That is to say, he mobilizes precisely those forms of visuality which inhere in a technology mapping a reality that otherwise eludes the naked eye. Chronophotography, in short, serves as a model for the veracity of the human eye.

The account of Saint-Loup's fistfight is therefore more than a mere exercise in cubist or futurist styles; it is a tribute to machines of vision. Like the depiction of Albertine's face, the 1907 piece on motoring, the Martinville episode, and the visually advanced description of the car ride up the hill to La Raspelière, the depiction of the fistfight draws on instrumentalized modes of vision, thus relying as much on induction as on deduction. In his remarkable attempt to offer a didactics of innocent perception that ultimately aims to name the perceived anew, Proust thus turns to unexpected sources. He taps the energies of those photographic machines that devour the epistemic privileges of the five senses and go about their tasks with, as Marey suggested, astonishing precision.

Proust's aesthetic endeavors, as I have attempted to show, may thus usefully be thought together with the implications of Marey's chronophotographic work. To consider Proust and Marey in tandem is to recognize that the historical premises and effects of visual technologies such as chronophotography and cinematography are inscribed in Proust's artistic program from the start. Although the terms in which it is couched are a structural inversion of the terms of Marey's theoretical enterprise, both operate within more or less the same discursive bounds. Both participate in the rearrangement of the limits of the visible and the invisible and, by implication, in the redefinition of the limits of the realm of objective knowledge vis-à-vis that of the merely subjective.

"Custom hangs upon us with a weight/Heavy as frost, and deep almost as life," Wordsworth once wrote.[95] We have seen how Proust evokes the Ruskinian ideal of perceptual innocence, seeking to rescue the immediacy of sensory impressions from the domesticating effects of intellectual notions. We have also seen how Proust's style draws upon and interacts with technologically mediated modes of representation, in particular photographic modes of representing time, speed, and movement. Nowhere does this become as evident as in "Impressions de route en automobile." A piece of writing that proved essential to both the genesis and the narrative order of Remembrance, the Figaro article explores the motorist's visual experience of the surrounding landscape before it has been subjected to the dulling force of habit. A new space of representation has burst open, enabled by technologies of speed as well as by those photographic means of representation that articulate the modern syntax of space-time.

Seen from the viewpoint of the aesthetics of perception inherent in the Elstir portrait, the various pieces and episodes I have discussed—the *Figaro* article, the Martinville episode, the railway journey to Balbec, Saint-Loup's fistfight, and others—show Proust's style at its most successful. They also show that *Remembrance* incorporates the instrumentalized modalities of perception inherent in those technoscientific configurations that annex the traditional epistemic mandates of the human senses. All of which testifies to the historical complexity and richness of Proust's aesthetics—as does the fact that *Remembrance* also helps provoke the emergence of a spectator disposed to the new forms of aesthetic gratification that modernist art calls forth in the wake of the epistemological crisis of the senses, offering an education of the eye whose impulse to perceive the world anew is inseparable from the processes of instrumentalization it simultaneously attempts to undo.

At one extreme, Proust's novel is given over to a contemplative inquiry into times lost, exploring the role of the five senses in the search through the past. At the other extreme, the novel is fuelled by a passion for the present, a restless, voracious, and transformative passion that revolves not around memory but around sense perception as such. In this chapter, I have put the spotlight on a forward-looking and future-oriented writer with a hearty appetite for the modern world of technology and the new ways of seeing it affords. When Proust's novel comes to a close, the horse-driven carriages in the Bois de Boulogne have been replaced by automobiles; telephone calls are part of daily life; and airplanes are simply airplanes and have nothing in common with those creatures hovering above the treetops between wings of steel. The old world has aged beyond recognition, and the new is no longer new. Proust knew how to capture the arrival of the modern at precisely that short-lived moment when the distant past rubs against the absolutely new, when the world suddenly and unexpectedly appears as a swarm of hieroglyphs crying for decipherment.

4 The Aesthetics of Immediacy

Ulysses and the Autonomy of the Eye and the Ear

As a cow devours grass, so literary themes are devoured; devices fray and crumble.

VIKTOR SHKLOVSKY

In 1919, Man Ray completed an airbrush painting and decided to call it *Admiration of the Orchestrelle for the Cinematograph.*[1] The enigmatic title is literally part of the picture, written in the lower left-hand corner (figure 12). Taken together, the title and the painting form a picture puzzle, a hieroglyphics.[2] Two precisely rendered objects occupy the visual center of Man Ray's glossy painting: first, a brownish horn, a synecdoche for a gramophone; second, a brownish disc pierced by a red ray, alluding to the human eye as well as to the camera iris. Each of the two devices hovers above an oval light-refracting lens, just about connected to one another by a thin line. The shape of the horn, moreover, resembles that of the plate, only pinched in the middle and folded. It is as if the horn and the disc were doubles—a theme reflected in the twin lenses, the twin squares, and the nearly parallel lines that run through and divide the space of the painting.

Executed well before the advent of sound film, *Admiration of the Orchestrelle for the Cinematograph* can be seen as a figuration of how cinematic sound, specifically music, approximates cinematic vision, thereby also thematizing the technologically mediated relation of the habits of the ear and those of the eye. Early cinema was a medium of cultural production whose capacity for analogically reproducing the real as it moved through time stood in sharp contrast to its other tendency: the compart-

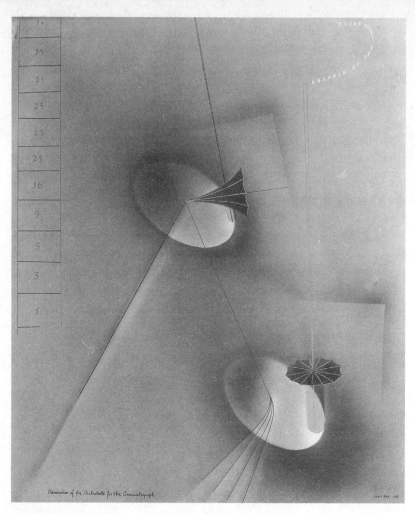

12. Man Ray, *Admiration of the Orchestrelle for the Cinematograph*, 1919. Ink and gouache airbrushed on buff paper, with traces of pencil. Museum of Modern Art, New York. Gift of A. Conger Goodyear. Photograph © 2001 The Museum of Modern Art, New York.

mentalization of the ways in which the five senses experience and process the real. If cinematography divorced sight from sound, it also relegated taste, smell, and touch to the margins of sense experience. Indeed, cinema, like photography half a century earlier, helped introduce a new optical space and reinforce a new perceptual division of labor. This time the photographic spectacle brought bodies in motion to the fore, except, of course, that the sounds in which motion was enveloped were absent. Just as humans laughed and babies cried without auditory traces, so trains and motorcars sped across the screen in utter silence.

Maxim Gorky, having attended the first screening of the Lumière program in Nizhni-Novgorod, detailed his bewildering experience of "moving photography" in a review article. "It is not life but its shadow, it is not motion but its soundless spectre." This was in July 1896, six months after the Lumières' historic show in Paris. Gorky referred to the screening room as the "Kingdom of Shadows," returning time and again to the uncanny nature of the soundless world spreading out before his eyes: "Noiselessly, the ashen-grey foliage of the trees sways in the wind, and the grey silhouettes of the people, as though condemned to eternal silence and cruelly punished by being deprived of all the colours of life, glide noiselessly along the grey ground."[3]

If cinematography helped invent new forms of visuality, it also helped invent new forms of audibility. Gorky's observations suggest that precisely because film was perceived as an unprecedented analogical mimesis of movement through time, it was all the more notable that it had been stripped of its acoustics. And if the advent of photography failed to elicit a similar reaction, it was partially because a long history of pictorial representation furnished viewing skills that could be applied to photographic images. Cinematic modes of representation, as Gorky's article intimates, called for new viewing skills, at the same time revealing, retrospectively as it were, that the reading of the "real"—spatiotemporal movement, for example—involved the processing of sounds. Thus, to apprehend the absence of sound is also to rediscover sound, in effect to reinvent it—in its pure and abstract form. The notion of pure and abstract sound that emerges in this historical period, more powerfully so than ever before, is best understood and described as overdetermined. It is mediated not only by the cinematic apparatus in the broadest possible sense, but by a diverse array of technological phenomena, among them devices for transmitting and reproducing acoustic data, such as the phonograph and the telephone.

Few literary works are as implicated in this social process, and in such intriguing and complicated ways, as James Joyce's *Ulysses* (1922). Indeed, as I shall argue in this chapter, *Ulysses* is a modernist monument to the eye and the ear. Plotting their uses, tasks, and internal configurations in the second machine age, Joyce's novel is both an index and enactment of the increasing differentiation of the senses, particularly sight and hearing. For this reason, *Ulysses* is also a monument to the technological changes that Proust's *Remembrance of Things Past* and Mann's *Magic Mountain* register, thematize, and explore, except in Joyce these changes have for the most part migrated into questions of form, more fully and radically so than in either Proust or Mann. This is to say that the historical processes that Proust's and Mann's novels subject to discussion are already sedimented in the narrative form of *Ulysses*.

Joyce himself, in a conversation with his friend Frank Budgen, said that his novel was, among other things, "the epic of the human body."[4] This was in Zürich during the Great War, while the Irish exile was working on *Ulysses*. Joyce wanted each episode of the epic to correspond to a particular bodily organ, from the brain to the sexual organs, including the eye and the ear. Taken together, the episodes would form an encyclopedia of the human body, thus giving shape to and amplifying what Joyce also referred to as a "little story of a day."[5] Joyce listed the various organs in a scheme, long since famous; and of the eighteen chapters, fifteen correspond to a specific organ. The other structuring devices included in Joyce's chart were the projected parallels in Homer's *Odyssey*, the hours of the day, the arts and sciences, colors, symbols, and techniques, all of which formed a vast set of correspondences.

Clearly, the function of Joyce's interpretive grid is analogous to that of the leitmotiv in Mann's *Magic Mountain;* both work to mold the reader's perception and comprehension of the narrative raw materials. Joyce's scheme, incidentally, began circulating well before the book itself was published, thus helping to engender a critical apparatus whose interpretive terms were largely resonant with Joyce's own intentions.[6] The most important parallel, in Joyce's view, was Homer's epic. Joyce thought of *The Odyssey* as a universal echoing chamber; and the Dublin adventures of his emphatically ordinary heroes, Leopold Bloom, Stephen Dedalus, and Molly Bloom, all drawn from the Irish middle classes, were ultimately to reflect the adventures of Odysseus, Telemachus, and Penelope, thus mirroring the ostensibly timeless quest at the heart of the ancient Greek myth.

Now, in projecting the idea of an epic of the human body, Joyce had more in mind than a merely formal scheme of somatic correspondences. "In my book," he explained, "the body lives in and moves through space and is the home of a full human personality. The words I write are adapted to express first one of its functions then another. In *Lestrygonians* the stomach dominates and the rhythm of the episode is that of the peristaltic movement."[7] Considered on Joyce's own terms, *Ulysses* might well be seen as a natural history of the body, a corporeal history that has been shrunk into some twenty hours in the life of two men and one woman, rendering all kinds of bodily action, including strolling, dying, eating, looking, shitting, digesting, drinking, sleeping, masturbating, and listening. This, of course, accounts for one of the greatly innovative, as well as greatly scandalous, features of Joyce's 1922 epic.

Apart from projecting that natural history of the body, however, Joyce's tale is also an allegory of a social history of the body. Indeed, the burden of this chapter will be to demonstrate that *Ulysses* registers the social history of that interface between world and embodied individual known as the sensorium. Joyce's epic pushes the physiological infrastructure and perception itself to the fore, far more resolutely and acutely than in either Mann or Proust. The act and means of sensory perception are not only foregrounded in Joyce's *Ulysses;* they have become problems in their own right, as have the various economies of stimuli the senses strive vigorously to process.

The human sensorium implied in *Ulysses* is no longer primarily associated with the attempt to elaborate a phenomenology of pure perception and corresponding forms of knowledge, as is the case in Proust. Nor is the Joycean obsession with the sensorium linked to an inquiry into the relativization of the epistemic mandates of the human eye, as is the case in Mann. Where Mann thematizes the division of labor between *aisthesis* and *theoria,* or the workings of the eye and the limits of what it may see and know, and where Proust explores ways of appropriating the perceptual modalities introduced by those vision machines that displace the epistemic mandates of the human eye, Joyce's self-styled experimental epic suggests that a certain historical process is more or less complete: each sensory organ now appears to operate independently and for its own sake. In fact, each sensory organ, particularly the eye, tends to perform according to its own autonomous rationality, as though detached from any general epistemic tasks. "His gaze," Joyce writes accordingly, "turned at once but slowly from J. J. O'Molloy's towards Stephen's face and then bent at once to the ground, seeking" (*U* 7.819–20).[8] Here the trivial activity of looking is rewritten as an event in itself. In Joyce, to look is no longer a mere predicate to be attached to a subject; the predicate has been unhinged from the subject and operates independently, endowed with an agency all its own.

In *Ulysses,* the workings of the entire sensorium are specialized and orchestrated with unprecedented intensity. When Joyce's hero, the ad salesman Leopold Bloom, stands face to face with a bookseller, the odor issuing from the latter's oral cavity emerges with graphic precision, as though the harsh smell were an alien object lodged in his throat: "Onions of his breath came across the counter out of his ruined mouth" (*U* 10.596–97). Thus Joyce's mode of figuration brings forth what is never more than implied: the smell, including Bloom's overwhelmed olfactory organ. As Joyce himself declared, "I want the reader to understand always through suggestion rather than direct statement."[9] Interestingly enough, Joyce's means of suggestion are frequently visual: that olfactory experience, for example, is translated into a visual impression, into something for the reader to *see*— which testifies to the inherent difficulty in rendering bodily sensations by way of language. This is one of the representational problems to which *Ulysses* attempts a solution.

That same morning, on June 16, 1904, Bloom had walked out of his house to purchase a few breakfast items, and before he knew it, he was arrested by the Polish butcher's tempting display. Joyce writes: "The shiny links, packed with forcemeat, fed his gaze and he breathed in tranquilly the lukewarm breath of cooked spicy pigs' blood" (*U* 4.142–45). Bloom is hungry, to be sure, but the mundanity of the scene should not prevent us from recognizing that the activity of the glossy sausages—their "feeding" Bloom's gaze—is not just a fortuitous metaphor. The sentence is an example of perfect symmetry; the sausages act on Bloom's sense of sight, just as Bloom's nose detects how they breathe. In other words, those links are no more, but also no less, an agent than Bloom himself. Sausages, onions, and the like make up a vast sensory space crowded by animated objects that actively enter into dialogue with Bloom's sensory apparatus, which they simultaneously help to define.

Yet although Joyce's novel invigorates the worlds of taste, smell, and touch, it is the remaining two senses that are singled out and privileged. Sight and hearing define the sensory universe of *Ulysses,* mediating the relationship between Joyce's subject-world and the corresponding object-world. But if sight and hearing reign supreme, their particular articulations and thematic clusters belong to this side of those historical breaks produced by the emergence of technologies for reproducing the visual and the audible respectively. In effect, a great number of Joyce's most characteristic and innovative stylistic strategies, among them representations of the human sensorium and modes of sensory activity (including their objects), must be thought of in the light of late nineteenth-century technological configurations such as cinematography, phonography, and telephony.[10]

I therefore propose to read *Ulysses* less as an epic of the human body than as a record of the modernist reinvention of the human body, particularly of the ways in which sensory habits are reconfigured, both internally and in relation to one another. By focusing on how the physiological interior mediates the exterior and vice versa, I will attempt to historicize a crucial, yet commonly neglected, part of the stylistic fabric that constitutes that "little story of a day": the representation of how Joyce's protagonists live their bodies and negotiate the sensuous world.[11] Despite the stylistic variegation that characterizes *Ulysses*, this set of features persists throughout the novel.

Fittingly enough, the narrative centers on a writer-to-be, Stephen Dedalus, and another equally sign-sensitive male protagonist, the ad canvasser Leopold Bloom (who, incidentally, at times appears as a frustrated writer). Both are concerned with the production and decoding of signs in a crowded urban milieu that is rendered as continuously supplying sensory stimuli.[12] At the same time, however, Stephen and Bloom are to be seen as formal units, as functions of the various modes of perception mapped out in *Ulysses*. In other words, by operating as what Viktor Shklovsky once called "connective tissue," they permit the novel's specific sensory worlds to come into being.[13]

The female protagonists are another story, at least insofar as modes of visual perception are concerned. The female characters map out a different world, as they tend to negotiate the real in more "organic" ways, which are, according to the novel's inherent aesthetic theory, less interesting ways. Gerty MacDowell, to take the most obvious example, defines a premodernist visual perspective on the world; for this reason, her mode of looking emerges in sharp contrast to those of Bloom and Stephen. Such a sexual division of labor, as I will discuss, throws a new light on the aesthetic program implicit in *Ulysses*.

In the following, I will trace two separate yet interrelated thematic paths: that of the ear and that of the eye. By way of conclusion, I will dwell on the ideal of the synaesthetic work of art and how it manifests itself in *Ulysses*. Two dominant formal tendencies may be observed in Joyce's novel, and they pull in opposite directions. The ideal of the organic form means that the parts make up the whole. Yet—and this incompatibility lies at the heart of *Ulysses*—the parts are not reducible to the whole. To explore how acts of perception are staged, including their implied matrices of perception, is to uncover a powerful stylistic tendency that, on the level of the sentence, pulls toward differentiation and autonomization. The ever-potential whole thus yields to the proliferation of details. This helps explain why, as Joseph Frank suggests in a now-classic phrase, "Joyce cannot be read—he can only be reread."[14] In effect, Joyce's writing is determined by this antinomy between part and whole, detail and totality, the general and the particular.

I. The Particular

The opening of *Ulysses* offers an everyday scene that unfolds before the reader in striking visual detail. Beginning *in medias res,* the opening pages dwell on how Stephen Dedalus and his two friends Buck Mulligan and Haines rise, chat, and have breakfast in the Martello tower. The first sentence introduces a perky Buck Mulligan and how he, "stately" and "plump," comes down the staircase. Wearing a yellow dressing gown that flutters round his body like a priestly mantle, he greets his half-awake friends with loud cries. A few sentences later, Stephen Dedalus enters the scenario, and at the same time, Joyce introduces a characteristic stylistic device, a trademark visualizing technique that, in various ways and with varying intensity, will be deployed throughout the eighteen episodes of *Ulysses.* This is how the implicit narrator details Stephen's visual perception of Buck Mulligan: "Stephen Dedalus, displeased and sleepy, leaned his arms on the top of the staircase and looked coldly at the shaking gurgling face that blessed him, equine in its length, and at the light untonsured hair, grained and hued like pale oak" (*U* 1.13–16).

Within the space of a few paragraphs, the visual representation of Mulligan, whom we just observed proceeding from the stairhead and into the room as though in a full-length portrait, has shrunk to a face. In fact, his face has been turned into a thing that takes on a life of its own. The horselike face is said to shake and gurgle all by itself, even bless a somewhat irritated Stephen. Dehumanized and reified, Mulligan's face floats like a hairy oval before the reader.

Subsequently, Mulligan brings his shaving utensils to the parapet, lathers his cheeks and chin, and begins to shave, meanwhile chatting with Stephen. Mulligan playfully raises the mirror to reflect a few sun rays, his cheerfulness focalized through Stephen: "His curling shaven lips laughed and the edges of his white glittering teeth. Laughter seized all his strong wellknit trunk" (*U* 1.131–33). Significantly, Joyce does not write that Mulligan is laughing, but that his lips are; likewise, Mulligan is not seized by laughter, but his stomach is. Joyce represents Mulligan's body—that is to say, his lips, teeth, and torso—as responding to external stimuli as though its reactions were mere reflexes, bypassing the control of some centrally operating intentionality. Mulligan's physical appearance turns into a miniature spectacle before the reader.

The aesthetic effect of such passages, so common in Joyce, depends on the differentiation of the human body, whose various parts are then autonomized and, furthermore, endowed with an agency all their own. In this introductory episode, as everywhere else in *Ulysses,* Joyce's implicit narrator builds on a narratological aesthetic that aims at defamiliarization. The narrator, one could say, keeps to what he perceives, not to what he knows is there. In this way, Joyce's aesthetic reveals deep affinities with that of Proust, although Joyce, as we shall see, will push that program to an extreme.

The Differentiation of the Sensory

When Mulligan is about to descend into the tower, leaving Stephen to ruminate over his dead mother, Stephen's visual perception of his roommate's bodily movement is rendered as it presents itself to his eyes. Buck Mulligan's figure thus appears as an optical outline, temporarily frozen by the entrance frame through which he is disappearing:

> His head halted again for a moment at the top of the staircase, level with the roof:
> —Don't mope over it all day, he said. I'm inconsequent. Give up the moody brooding.
> His head vanished but the drone of his descending voice boomed out of the stairhead. (*U* 1.233–38)

From a visual point of view, Mulligan's bodily whole has been bisected by the frame through which he passes. All Stephen perceives is a head, as though Mulligan's body had been sliced into two parts. There is a striking affinity between Stephen's image and a photographic frame, that instant freezing of time and movement. From a rhetorical point of view, Mulligan's visual gestalt has been substituted for a synecdoche, his thinglike head being the sign that stands in for the whole and whose shape can be observed for a few more moments. As Alan Spiegel argues, such a stylistic method—to render wholes by their parts—is "the characteristic formal procedure of Joyce's modernism." Furthermore, Spiegel relates this stylistic feature to cinematic techniques.[15] But what, exactly, is the whole, the gestalt? The passage suggests that Stephen's perceptual experience of Mulligan's descent is processed in two different registers. On the one hand there is Stephen's visual impression, and on the other, the auditory one. Each is distinct; indeed, each is separate and independent of the other:

Buck Mulligan's voice sang from within the tower. It came nearer up the staircase, calling again. Stephen, still trembling at his soul's cry, heard warm running sunlight and in the air behind him friendly words.

—Dedalus, come down, like a good mosey. Breakfast is ready. Haines is apologising for waking us last night. It's all right.
—I'm coming, Stephen said, turning.
—Do, for Jesus' sake, Buck Mulligan said. For my sake and for all our sakes.
His head disappeared and reappeared. (*U* 1.281–89)

What is heard is not joined together with what is seen; and what is seen is in its turn a mere slice of the whole. The multisensory hermeneutic horizon, the all-embracing gestalt, refuses to take shape. But since Joyce, in this passage and elsewhere, aligns himself with a modernist aesthetic that aims to render what is perceived rather than what is known, traditional ways of describing movement, gestures, and action necessarily have to be challenged, and with them, the idea of "organic" modes of perception.

Mulligan's descent into the tower not only suggests how Stephen's habits of perception are differentiated; it also expresses the tendency to which such differentiation is dialectically related: synaesthesia. For once the work of the senses has been separated, as this passage intimates, new perceptual and descriptive possibilities open up, and now the perceptual activity of a single sense may well be translated into that of another, even give way to a synaesthetic experience: "Stephen, still trembling at his soul's cry, heard warm running sunlight." Thus Stephen not only hears the sunlight; his aural impression of the morning light is also refracted through categories of tactility: the running rays are warm to the touch as they spread across Dublin.

The Autonomy of the Ear

Joyce's mode of representing Stephen's sensory impressions in the Martello tower scene—the disembodied and thinglike voice rising out of the tower, Buck Mulligan's decapitated head appearing and disappearing, and so on—are refracted through a perceptual matrix enabled by technologies for transmitting and reproducing the real, acoustic and visual technologies alike. Indeed, such a pronounced desire to represent what is heard and, furthermore, to represent it in a register that is radically separate from what is seen, must be considered in light of those late nineteenth-century acoustic technologies that mediate the new matrices of perception, readily turning the sense of sight and that of hearing into autonomous activities. For once we place Joyce's mode of representing aural impressions alongside the theory of sensory reification embedded in Proust's telephone episode, we realize the great extent to which the very experiential effects that Proust's narrator contemplates ("a tiny sound, an abstract sound") effectively precede and inscribe some of the most persistent and characteristic stylistic aspects of *Ulysses*.[16] A similar pattern, incidentally, may be distinguished in Woolf's 1937 novel *The Years*, in which the telephone figures largely and human voices are emphatically perceived as autonomous entities, on or off the phone: " 'Hullo,' said a gruff voice, which suggested sawdust and a shelter, and she gave the address and put down the telephone."[17]

As for the dissociation of the eye and the ear in Joyce's novel, one further example will suffice.[18] Stephen is in the classroom teaching his rather unwilling students history, and suddenly they are alerted to a sound:

A stick struck the door and a voice in the corridor called:
—Hockey!
They broke asunder, sidling out of their benches, leaping them. Quickly they were gone and from the lumberroom came the rattle of sticks and clamour of their boots and tongues. (*U* 2.118–22)

More than anything else, this stylistically sophisticated miniature scene serves to characterize Stephen's sensory apparatus. Beginning with a voice stripped of its author, the passage proceeds to render the students' sudden movements in all their visual purity, only to close with sounds, more specifically, with the acoustic phenomena issuing from the lumberroom. Again there is symmetry between the human and nonhuman; in typical spirit, the Joycean narrator levels the sound of thundering tongues with the sound of thundering boots, as though no particular distinction is to be made between human and nonhuman sounds. Stephen stays behind with one of the students, Sargent, who needs extra assistance, until

In the corridor his name was heard, called from the playfield.
—Sargent!
—Run on, Stephen said. Mr Deasy is calling you.
He stood in the porch and watched the laggard hurry towards the scrappy field where sharp voices were in strife. [. . .]

Their sharp voices cried about [Mr Deasy] on all sides: their many forms closed round him, the garish sunshine bleaching the honey of his illdyed head. (*U* 2.181–98)

Once again the voice is represented as a thinglike and autonomous entity.[19] The boys' voices act on their own, as though bypassing screens such as the cortex, spreading their sharp vibrations all the way to the veranda where Stephen is standing. Meanwhile his visual impression of the boys' appearances fades. Gradually, they blend into so many optical outlines surrounding the stingy headmaster whose hair color stands out as a sunny exclamation mark. Mr Deasy's appearance is similarly focalized through Stephen when, a few pages later, we are told that "[Mr Deasy] raised his forefinger and beat the air oldly before his voice spoke" (*U* 2.345). To Stephen, Mr Deasy appears as something like an automaton, as do numerous people who pass his field of vision.

These passages are drawn from "Nestor," the second episode in *Ulysses.* Episodes one through three all focus on Stephen Dedalus and his morning adventures (it is only in the fourth that Leopold Bloom appears on Joyce's stage); and in the third, "Proteus," which I will discuss shortly, Stephen's introduction into the sensory world mapped out by *Ulysses* reaches its climax. For here, fittingly enough, Joyce has Stephen ponder the nature of perception itself, specifically the way in which the senses mediate the experience of the world. Of the five senses, Stephen singles out two: seeing and hearing.

As we shall see, Bloom also moves through the world using his eyes and ears, and his sensory organs, too, tend to operate separately. The dissociation of the visual and the aural runs through Joyce's novel from beginning to end, organizing its many modes of representation and stylistic experiments.[20]

The Autonomy of the Eye

That Joyce was musically oriented and that *Ulysses* in many ways is a musically inflected composition are commonplaces in Joyce criticism.[21] It is true that acoustics, music, noise, rhymes, assonances, alliterations, and the mimesis of sounds play a far more important role in Joyce's encyclopedic novel than in, for example, Mann or Proust.[22] This critical focus on the aural and the musical has nevertheless tended to obscure the crucial role of vision and visuality in *Ulysses*. For visuality prevails in *Ulysses,* and it does so in a variety of ways.[23] At the same time, however, I want to argue that in Joyce—and, ultimately, in modernism at large—forms of visibility and forms of audibility have to be considered in tandem. Mutually defining one another, the abstraction of the seen mediates the abstraction of the heard, and vice versa.

To begin on the most obvious level, that of motifs and description, *Ulysses* displays an obsessive fascination with eyes, a preoccupation that reverberates in numerous modernist novels, notably, as we have seen, Mann's *Magic Mountain*. On virtually every page in *Ulysses*, whether in "Nausicaa," designated by Joyce as the eye and nose chapter, or in "Sirens," the ear episode, or in "Penelope," the flesh chapter, there are numerous references to eyes.

Similar tendencies may be observed in, for example, Joseph Conrad's *Lord Jim* and Wyndham Lewis's *Tarr*, both of which fixate on eyes, practically making them into characters in their own right. This is how Lewis describes the tension between two strangers in a restaurant, a charged specular drama performed by two pairs of eyes: "He had already been examined by the beautiful girl. Throwing an absent far-away look into her eyes, she let them wander over him. Afterwards she cast them down into her soup. As a pickpocket, after brisk work in a crowd, hurries home to examine and evaluate his spoil, so she then examined collectedly what her dreamy eyes had noted."[24]

As Lewis's vibrant imagery suggests, once the devouring eye has been autonomized and endowed with agency, new representational realms open up and the human body readily transforms into spectacle, including the eye itself. Robert Musil, for one, likens his hero's gaze to an insect buzzing around metropolitan space.[25] Similarly, Lewis superimposes a new scenario on the initial one. As the organ of sight is turned into a petty thief engaged in criminal acts, the unexpected imagery sheds new light on the entire episode, now brimming with all the breathless excitement that derives from stealing a look at something one desires but is afraid to acknowledge.

Joyce's 1916 novel *A Portrait of the Artist as a Young Man* also displays a fascination with eyes. One scene in particular echoes the visual syntax in Lewis's elaborated passage above. Joyce's male artist-hero is strolling by the sea and catches sight of a girl. Voraciously consuming her appearance, Stephen's reifying gaze divides her physique into parts, beginning with her soft, ivory-hued, and cranelike legs, moving on to her delicious bosom and finishing with her beautiful face. The girl then looks at Stephen: "She was alone and still, gazing out to sea; and when she felt his presence and the worship of his eyes her eyes turned to him in quiet sufferance of his gaze, without shame or wantonness. Long, long she suffered his gaze and then quietly withdrew her eyes from his and bent them towards the stream, gently stirring the water with her foot hither and thither."[26]

It is no coincidence that in both Lewis and Joyce the gazer is male and the gazed-upon female. In modernism at large, the alienating and reifying gaze is usually associated with masculinity, and as we shall see, this sexual division of labor is particularly striking in Joyce's *Ulysses*, indeed a vital component of its aesthetics of perception.

If there is one attribute Joyce remarks on when describing his characters, it is their eyes. In *Ulysses*, the protagonists' eyes are frequently represented as signifying a personality trait or emotional state, temporary or permanent. There are, for instance, Buck Mulligan's "smokeblue mobile eyes" and, alternatively, his "pale Galilean eyes." Mr Deasy, for his part, is endowed with "seacold eyes." Or think of Mr Kernan's "bloodshot eyes. Secret eyes, secretsearching." Further, Mrs Breen has "womaneyes," just as Mr John Eglinton is associated with "miscreant eyes glinting." An old shopman's gaze bespeaks disease, with his "eyes bleared with old rheum." The reader also encounters John Fanning's "large fierce eyes"; John Wyse Nolan's "cool unfriendly eyes"; John Henry Menton's "winebig oyster eyes"; Boylan's "spellbound eyes"; and Miss Douce's "brave eyes, [. . .] smitten by sunlight." The examples could be multiplied.

What is more important, however, is that eyes are so frequently endowed with agency. Eyes do things. Eyes even participate in the plot, most obviously in "Nausicaa" where Leopold Bloom and Gerty MacDowell enjoy a momentary sexual encounter of sorts by exchanging looks. In effect, eyes in *Ulysses* emerge as characters in their own right, appearing to operate independently of the human consciousness to which vision nevertheless is related. Eyes claim autonomy for themselves, not just from the other senses and the human body at large but also from a central processing instance, the hermeneutic switchboard called the brain. This, for example, is how Joyce represents the visual faculty of Mr Deasy, Stephen's superior: "His eyes open wide in vision stared sternly across the sunbeam in which he halted" (*U* 2.357–58). In its context such a description, focalized as it is through Stephen, works to estrange its object, underscoring Stephen's skepticism toward the headmaster. Yet such stylistic techniques abound in *Ulysses*, and their function is to suggest how the heroes themselves, Stephen Dedalus and Leopold Bloom, perceive and interact with the world, its human and nonhuman components alike. Thus Bloom's gaze, too, acts on its own, as though detached from a conscious nucleus: "His eyelids sank quietly often as he walked in happy warmth" (*U* 4.81);

"His eyes rested on her vigorous hips" (*U* 4.148); "His gaze passed over the glazed apples serried on her stand" (*U* 8.70); "His eyes sought answer from the river and saw a rowboat rock at anchor on the treacly swells lazily its plastered board" (*U* 8.88–89); "His oyster eyes star[ed] at the postcard" (*U* 8.322).

Given the implied self-sufficiency of Bloom's and others' eyes, it is appropriate that it takes an effort to control their wanderings, as when Bloom's eyes are drawn to a woman's hips: "He held the page aslant patiently, bending his senses and his will, his soft subject gaze at rest" (*U* 4.162–63). Or when Bloom pays for the kidney: "A speck of eager fire from foxeyes thanked him. He withdrew his gaze after an instant" (*U* 4.186–87). Or, finally, when Bloom is about to collect the secret love letter from his pen pal, Martha: "From the curbstone he darted a keen glance through the door of the postoffice" (*U* 5.52–53). Furthermore, once looking has become autonomous and is rendered as such, reification is close at hand, and a gaze may then appear as palpable an object as a beard: "From a long face a beard and gaze hung on a chessboard" (*U* 10.425).

Even if references to self-sufficient eyes outnumber those to any other organ or body part, nearly all body parts and extremities do things in *Ulysses*. Hands, fists, fingers, fingertips, thumbs, ears, feet, and tongues also perform. They do whatever they do separately and independently from one another, as though each disembodied part had been furnished with a consciousness of its own. A tongue, for example, may decide to join the activity of the eyes, setting out to reinforce what the eyes themselves are trying to do and say, as when Mr Bloom runs into Mrs Breen who tells him about poor Mina Purefoy, taken ill after a difficult childbirth: "His heavy pitying gaze absorbed her news. His tongue clacked in compassion. Dth! Dth!" (*U* 8.287–88) As for sudden animation, consider Professor MacHugh as he lectures eloquently on Israel's enslavement by the Egyptians: "A dumb belch of hunger cleft his speech. He lifted his voice above it boldly" (*U* 7.860–61).

Joyce's penchant for autonomization and animation is true not only of his descriptions of eyes and the human body but of the thing-world at large. The "Circe" episode is a case in point. There as elsewhere in *Ulysses*, inanimate objects have been upgraded to the level of human agency. The gramophone, the bells, the gong, the chimes, and the pianola, not to mention Bloom's lemony soap—all these objects play proper roles in the phantasmagoric drama, speaking, acting, singing, and otherwise interacting with the protagonists swirling around in the Nighttown quarters. In fact, the Joycean world in *Ulysses* has been subjected to such degrees of anthropomorphism and personification as to be almost leveled: "Quick warm sunlight came running from Berkeley road, swiftly, in slim sandals, along the brightening footpath" (*U* 4.240–41). As I suggested apropos of Bloom's animated kidney universe, humans and human bodies appear to have no more—but also no less—agency than that shiny link that feeds his gaze.

As for Joyce's mode of representing human agency and intentionality in such a leveled universe, here is Bloom a little later that warm June morning, standing in front of an ad:

> While his eyes still read blandly he took off his hat quietly inhaling his hairoil and sent his right hand with slow grace over his brow and hair. Very warm morning. Under their dropped lids his eyes found the tiny bow of the leather headband inside his high grade ha. Just there. His right hand came down into the bowl of his hat. His fingers found quickly a card behind the headband and transferred it to his waistcoat pocket.
>
> So warm. His right hand once more more slowly went over his brow and hair. Then he put on his hat again. (*U* 5.20–28)

Very little appears to happen here. A man looks at a poster, removes his hat, wipes his forehead, finds the secret postcard he has hidden inside the hat, places it in a pocket, and then puts on the hat again. Yet as soon as the passage is rewritten in such standard anthropomorphic terms, Joyce's stylistics loses its signal freshness. Karen Lawrence has analyzed the defamiliarizing tendency in Joyce, noting that "the description of certain common events like a handshake, a sunrise, a bump on the head, are documented with such precision that they are almost unrecognizable."[27] Indeed, a trivial activity such as taking off one's hat and wiping off sweat can be presented as an event of the first rank only when it has been defamiliarized, that is, when the languages and conceptual figures commonly used to describe such gestures are subjected to estrangement and thus revealed. This is achieved when Bloom's two eyes, his right hand, and his fingers all operate on their own, yet in concert; when, in other words, Bloom's body appears as an assemblage of independently operating parts, each differentiated, autonomous, and functionalized. As a result, the notion of a conscious nucleus is displaced, and a dissociated Bloom stands before the reader, reborn in each new visual frame which presents the physiological makeup of the hero in what could be described as a series of close-ups.

Similarly, when Bloom is about to purchase the breakfast item he has selected at Dlugacz's, the physical activities involved are broken down into what looks like their smallest component parts: "His hand accepted the moist tender gland and slid it into a sidepocket. Then it fetched up three coins from his trousers' pocket and laid them on the rubber prickles. They lay, were read quickly and quickly slid, disc by disc, into the till" (*U* 4.181–84).

In sum, Bloom pays for the kidney. His dismembered and self-sufficient hand could easily have passed as a synecdoche, as a part standing for the absent yet implied whole, had not the subsequent sentence—"*it* fetched up three coins"—picked up and amplified the syntactic tendency, underscoring that Joyce has a radically different kind of figuration in mind, that is, reification. The everyday business between customer and seller emerges as an almost automatic and at any rate anonymous exchange, a well-rehearsed miniature drama performed by a lonely yet target-oriented hand, a pair of eyes reading the coins and yet another hand, though invisible, sliding the coins into the cash register.

Joyce's appetite for the most common of common activities—taking off one's hat, wiping one's forehead, grocery shopping, and so forth—has a certain affinity with Proust's devoting some thirty pages to his narrator's initial turning round in bed. But while Proust's bedtime digressions motivate miniature essays on the psychology of habit, human consciousness, and architecture, Joyce's representation of an everyday man going about his everyday kidney-shopping serves no purpose except, as Conrad exhorted, "to make you *see*."[28] That is, while Proust's narrator turns mundane scenes and common incidents into occasions for deriving new forms of knowledge, thereby also treating them as open to a certain *generalization,* for Joyce the general, universal, and representative are anathema. While it is in the nature of the notion of the everyday to be based in the general, the point is that Joyce represents the everyday so as to make it utterly particular and immediate, hence immune to generalization. Indeed, the meticulously rendered scene at the butcher's cash register adds little, if anything, to the reader's understanding of Bloom's psychic profile, let alone the diegesis of the narrative.

What Shklovsky observed apropos of the function of Tolstoy's stylistic devices is therefore partly true of Joyce's descriptive techniques as well: "he does not call a thing by its name, that is, he describes it as if it were perceived for the first time, while an incident is described as if it were happening for the first time."[29] In other words, for such stylistic techniques to become effective, there have to be common names and habitual denominations already in place. Just as the notion of the everyday implies a recognizable generality, so it is in the nature of the particular that it resists common names and, by the same token, its estrangement. What distinguishes *Ulysses* is that its subject matter is the ordinary and little else. And this in the most profound sense: *Ulysses* grapples with the everyday in all its irreducible minutiae. At the same time it sets itself a taboo: a taboo against the case study. A question then arises. How is the particular to resist the stamp of the general?

Because Joyce's aesthetics is an aesthetics of immediacy, of the unmediated, the everyday has to be named anew, and continually, in order to retain its desired immediacy. Only when particularized can the common preserve its immediacy, and this is why, in Joyce, the imperative to make you *see* is so often an aesthetic end in itself. Seeing for the sake of seeing thus acquires a value on its own, and more radically so than in, for example, Conrad, Lewis, Proust, or Mann. Joyce's figurations are deliberately and emphatically dissociated from the "meanings" they simultaneously help to create and sustain. And even if the reader reconstructs possible references—Bloom is sweaty, Bloom pays for the kidney, and so on—the visual detail will always triumph over the progress of the plot, just as the particular refuses to merge with the whole.

Consider, for example, a passage in "Hades." Four men, among them Bloom, are on their way to Paddy Dignam's funeral. Seated in an old funeral carriage, they travel slowly through the streets of Dublin. The window of the carriage works as a frame, through which the men, along with the reader, look out. What they see is presented as though in a photographic shot, as if a camera had zeroed in on and framed the objects, placing them at the center of the delimited visual field: "All watched awhile through their windows caps and hats lifted by passers. Respect" (*U* 6.37–38). Here, the gestures of the passersby and their significance have been disconnected at the outset. The single word "respect" works to associate the sensuous signifier with the implied signified, suggesting how the four mourners perceive and, subsequently, read what they see: dismembered arms and raised hats, thus tributes to the deceased Paddy Dignam. The proper content of the passage, then, is the mourners' processing of a number of sense data, as though the implicit narrator endeavors to report what they actually see, not what they know is there.

Both Proust and Joyce can be said to pursue something like a phenomenology of pure perception. Yet where Proust, in order to render the freshness of the seen, most often has recourse to cinematic representations of motion and speed, Joyce tends to adapt cinematic framing techniques.[30] The "Aeolus" episode offers a rich example. When Bloom stops by his office to check on an advertisement, a colleague taps on his arm, bringing to his attention the entrance of a significant person, and this is how Bloom's view is represented: "Mr Bloom turned and saw the liveried porter raise his lettered cap as a stately figure entered between the newsboards of the *Weekly Freeman and National Press* and the *Freeman's Journal and National Press*. Dullthudding Guinness's barrels. It passed statelily up the staircase, steered by an umbrella, a solemn beardframed face. The broadcloth back ascended each step: back" (*U* 7.42–47).

The person in question is William Brayden, the owner of the papers. To Bloom, Brayden is a mere automaton. Defamiliarized and reified by Bloom's mocking gaze, the pompous owner turns into a neutered figure, indeed into a desexualized thing ("it"), then into a face, and finally into a back. Also, it is fitting that Brayden's movement should be driven by something as banal as an umbrella, that vulnerable phallic supplement and class attribute. From a technical point of view, the passage suggests how Brayden's continuous ascending presents itself to Bloom's and his colleague's field of vision as framed by the door opening. First they see a face, and then a back, two subsequent frames of frozen motion. The owner's physique finally disappears out of sight, and those remaining split seconds of upward motion are represented in three successive frames: "They watched the knees, legs, boots vanish" (*U* 7.63). But this analysis of Bloom's mode of vision may also be turned around; it is the door frame that permits and, indeed, naturalizes Joyce's stylistic decomposition of the owner's moving body into so many distinct analytic moments.

Consider yet another scene, focalized through the drunken, one-legged sailor who jerks himself along Eccles Street while growling "For England . . . home and beauty." Here a window motivates the visual framing of the event. The singing sailor approaches the window: "The gay sweet chirping whistling within went on a bar or two, ceased. The blind of the window was drawn aside. A card *Unfurnished Apartments* slipped from the sash and fell. A plump bare generous arm shone, was seen, held forth from a white petticoatbodice and taut shiftstraps. A woman's hand flung forth a coin over the area railings. It fell on the path" (*U* 10.249–53).

A blind, a card, an arm, a hand, a coin: from the autonomy of the eye it is a short step to the autonomy of the seen. In such an optical space, seemingly insignificant objects and phenomena take center stage, as Molly's solitary spoon does when she suddenly smells that burning kidney. "Her spoon ceased to stir up the sugar. She gazed straight before her, inhaling through her arched nostrils" (*U* 4.378–79). Such modes of figuration express processes of autonomization and leveling all at once, and they are inscribed by a specifically cinematographic logic. Gilles Deleuze, in

Cinema 1, designates this logic as "dividual" (*dividuelle*) and, therefore, deterritorializing: "The cinematographic image is always dividual. This is because, in the final analysis, the screen, as the frame of frames, gives a common standard of measurement to things which do not have one— long shots of countryside and close-ups of the face, an astronomical system and a single drop of water—parts which do not have the same denominator of distance, relief or light. In all these senses the frame ensures a deterritorialization of the image."[31] And so, having achieved visual autonomy, Molly's teaspoon, Brayden's umbrella, a woman's plump arm, a passerby's lifted hat, or Bloom's right hand may enter into new relationships with objects in space. Framed, leveled, and deterritorialized, these diverse phenomena effectively turn into spectacles before the reader's eye.

Yet another passage suggests how framing techniques are at work in *Ulysses.* Bloom has boarded the tram and engages half-heartedly in a conversation with a colleague, meanwhile observing an elegant lady seated not far from him. A tram passes, then takes off in another direction, and Bloom again claps his eyes on the lady whose appearance turns into a proto-cinematic spectacle. "The tram passed. They drove off towards the Loop Line bridge, her rich gloved hand on the steel grip. Flicker, flicker: the laceflare of her hat in the sun: flicker, flick" (*U* 5.138–40).

Refracted by the moving tram's window frames, the sunlight flows intermittently and transforms the lady's hand and hat into dazzling objects for Bloom's gaze to devour. The window frames thus work like a projector apparatus or like a shutter. Interestingly, Austin Briggs has suggested that the "flickering" in the passage refers to the uneven shutter speed of early projectors which, because they were hand cranked, tended to produce a jerky movement on the screen.[32] (It is interesting to note, in passing, that in Proust speedy vehicles give rise to panoramic representations of the visual and not, as is the case in Joyce, tightly framed vision.) Naturalizing the deliciously fragmented representation of Bloom's visual activity, the window frames permit the spectacle. The tram, then, functions as a vehicle of perception; and the entire tram segment of the episode serves to characterize Bloom's visually dominated sensorium.[33]

László Moholy-Nagy, a perceptive admirer of *Ulysses,* once suggested that Joyce's was a language of unprecedented precision. Joyce's almost microscopic attention to detail, he maintains in *Vision in Motion* (1947), makes one see more. Moholy-Nagy makes analogies to cubist painting and cinematography, but also to new visual techniques for representing bodily organs. His example, typically, is a kidney, whose blood system could be visualized thanks to new techniques: "By injecting latex into the blood vessels and dissolving the tissue in acid after the rubber has set, an exact replica of the kidney can be produced with outside *and* inside visible. With this new technique, as with the new writing, *one sees more.*"[34] Joyce's language of precision and passion for detail similarly make us see more, Moholy-Nagy concludes, although he bypasses the question of the nature of this "more."

To take the argument a step further: Joyce's visuals do not really make us see more; they make us see differently. In the wake of cinematography and its accompanying matrices of perception, it is the modalities of visibility that change. In a similar vein, Walter Benjamin suggests that a different nature offers itself to the camera than to the human eye; he designates it as an optical unconscious.[35] Joyce thus represents the peak of what Italo Calvino has called "the crisis of visibility":

> There is a history of visibility in the novel—of the novel as the art of making persons and things *visible*—which coincides with some of the phases of the history of the novel itself, though not with all of them. From Madame de Lafayette to Benjamin Constant the novel explores the human mind with prodigious accuracy, but these pages are like closed shutters which prevent anything else from being seen. Visibility in the novel begins with Stendhal and Balzac, and reaches in Flaubert the ideal rapport between word and image (supreme economy with maximum effect). The crisis of visibility in the novel will begin about half a century later, coinciding with the advent of the cinema.[36]

Indeed, with cinema and related technologies of visual representation, a different optical vocabulary slides into place, a tendency that Joyce's language of precision both draws upon and reinforces. Sergei Eisenstein, who was passionately interested in *Ulysses* and lectured on the novel numerous times, even went as far as to claim that Joyce's modernist text provides an excellent point of departure for "training visual consciousness."[37]

The Modernist Conquest of the Visual: Léger, Vertov, Shklovsky

The hand, Fernand Léger remarked in an essay written in 1925, is more than just a hand. The visual meanings of the hand are multiple, and cinema is the proof: "Before I saw it in the cinema, I did not know what a hand was!"[38] Léger's writings on visual culture may usefully be read in tandem with Joyce's aesthetics of detail, as they throw Joyce's zest for the visual into historical relief. Spanning the years that Joyce was working on *Ulysses,* Léger's aesthetic has a family resemblance to that of Joyce. In a series of essays and manifestos published in the 1920s, the French painter and theorist promoted what he called "the machine aesthetic [*l'esthétique de la machine*]." The modern age, Léger maintained, inaugurated a new art; and at the center of this historical shift stood the new technologies, cinema in particular. Cinema helped level traditional hierarchies of beauty, inviting spectators to discover for themselves the everyday world of objects, domestic or industrial, hence also a new standard of beauty. "The Beautiful is everywhere," Léger suggested, "perhaps more in the arrangement of your saucepans in the white walls of your kitchen than in your eighteenth-century living room or in the official museums."[39]

Above all, he argued, cinema was capable of transforming habitual modes of visual perception. In awakening the dull eye, cinema sensitized humans to the visual immediacy of ordinary objects. Cinema, in short, turned the world into a ceaseless flow of spectacles. In Léger's view, a visual revolution was in the making:

> 80 percent of the elements and objects that help us to live are only noticed by us in our everyday lives, while 20 percent are *seen*. From this I deduce the cinematographic revolution is *to make us see everything that has been merely noticed*. Project those brand-new elements, and you have your tragedies, your comedies, on a plane that is uniquely visual and cinematographic. The dog that goes by in the street is only noticed. Projected on the screen, it is seen, so much so that the whole audience reacts to it as if it discovered the dog.
>
> The mere fact of projection of the image already defines the object, which becomes spectacle.[40]

Understood on their own terms, Léger's writings on visual culture proclaim a new painterly aesthetics, which, based upon the opposition between heightened and habituated perception, ultimately entails a transfiguration of the notion and status of art as such. This theoretical effort went hand in hand with the promotion of a more general sensibility whose task was to undo the work of perceptual habit, restore to freshness that unseen 80 percent, and so transform the everyday experience of the sensuous world. But Léger's reflections may also be considered as a theory of the modern and therefore a theory of history. It is a theory for which cinema and the visual arts serve as the privileged occasion. By virtue of theorizing the modern in terms of visuality, he locates visual perception at the very heart of the experience of the modern.

Léger's theoretical essays also anticipate his experimental film *Ballet mécanique* (1924), particularly its penchant for the animation of inanimate objects—consider, for instance, that pair of mannequin legs dancing mechanically in frame after frame. In many ways, Léger's film has affinities with Dziga Vertov's *The Man with a Movie Camera* (1929), not only because both indulge in prolonged close-ups of the human eye, but because both feature a propensity for turning the inanimate object into spectacle.[41] Vertov himself referred to his film as a "visual symphony."

Vertov, like Joyce before him, takes as his subject matter a day in the life of a modern city. Casting a spell over its human and nonhuman subjects alike, Vertov's camera sweeps an entire city into spirited movement. Chimneys, workers, typewriters, street crossings, automatons, cars, smiles, sewing machines, pedestrians, bicycles, stockings, streetcars, shop windows, telephones: all participate in Vertov's rapturous urban ballet, effectively canceling T. S. Eliot's Waste Land, that Unreal City with its flow of dead masses and gloomy typists playing the gramophone at the end of a mindless day. Meanwhile Vertov's urban space, utopian and visually exuberant, turns into a narrative character in its own right, vitalized by a new language—cinema—which triumphantly declares its independence from traditional art forms, notably literature and theater. As Vertov proclaimed, *The Man with a Movie Camera* "aims to create a truly international film-language, *absolute writing in film,* and the complete separation of cinema from theater and literature."[42]

Vertov offered a theory of this new mode of cultural production. In his view, documentary cinema could participate in revolutionary transformation, and Vertov designated the "kino-eye," or "film-eye," as one of its prime vehicles. By conquering the visual in all its previously invisible aspects, Vertov suggested, the kino-eye also articulates the visual: "Kino-eye is the documentary cinematic decoding of both the visible world and that which is invisible to the naked eye."[43] In fact, the kino-eye even redefines traditional categories of time and space, subsuming them under the absolute priority of the visual:

> Kino-eye means the conquest of space, the visual linkage of people throughout the entire world based on the continuous exchange of visible fact, of film-documents as opposed to the exchange of cinematic or theatrical presentations.
>
> Kino-eye means the conquest of time (the visual linkage of phenomena separated in time). Kino-eye is the possibility of seeing life processes in any temporal order or at any speed inaccessible to the human eye.[44]

Cinema, then, not only generates a new representation of the world, now emerging as an exclusively visual artifact. Cinema also reformulates the categories through which this world is to be perceived, apprehended, and processed. It is appropriate, therefore, that *The Man with a Movie Camera* should highlight its own aesthetic devices, continuously commenting upon its own mode of production and representation. Like Proust's *Remembrance*, Vertov's film incorporates within itself its own beginnings, while at the same time offering an implicit aesthetics with which to grasp the narrative materials. In Vertov, the ever-present film photographer, the rolling camera, shootings of shootings, film developing, the film that freezes into stills—all become part of the narrative content which is thereby thrown into relief. We know and recognize such an aesthetic under the rubric of defamiliarization, or *ostraniene*, coined by Shklovsky in 1917. On Shklovsky's theory, the task of art as such is to subvert habitual perception, or as he also calls it, "automatized" perception:

> In order to return sensation to our limbs, in order to make us feel objects, to make a stone feel stony, man has been given the tool of art. The purpose of art, then, is to lead us to a knowledge of a thing through the organ of sight instead of recognition. By "enstranging" [*sic*] objects and complicating form, the device of art makes perception long and "laborious." The perceptual process in art has a purpose all its own and ought to be extended to the fullest. *Art is a means of experiencing the process of creativity. The artifact itself is quite unimportant.*[45]

Clearly, Shklovsky's notion of the literary device, like Léger's theory of visual art, is based upon the opposition between heightened and habituated perception, or, in his terms, estrangement and recognition, sensation and numbness. In Shklovsky, as in Léger, the eye is to carry out the task of making strange, thus enabling the discovery of things as they are in and of themselves. The act of perceiving the stoniness of the stone, then, is an aesthetic end in itself. Because habituated perception equals "automatized" perception, it is easy to identify how yet another modernist contradistinction is at work in Shklovsky's aesthetic: that between the mechanical and the corporeal. As we have seen throughout the previous chapters, in accord with the logic of this specifically modernist discourse, art and aesthetics have to come down on the side of the corporeal. Indeed, aesthetics is located in the corporeal materiality of the senses, a materiality that, crucially, is posited as an utterly irreducible and unmediated entity.

Joyce, Léger, Vertov, Shklovsky: as this family resemblance suggests, they all articulate the primacy of the visual. In effect, Bloom's right hand—that Bloomy hand-in-itself—is to Léger's dancing mannequin leg as Shklovsky's stony stone is to Vertov's exuberant city machines. Bracketed from their common uses, habitual denominations, and conceptual figures, these phenomena are all visibly revealed—not so much recognized as invented anew—in their purely and emphatically visual aspects. Thus Joyce, Léger, Vertov, and Shklovsky all participate in the modernist conquest of the visual, embracing an emergent worldview in accord with which the world is increasingly managed by way of its aestheticization.

Stephen's Theory of the Senses

For a narrative that is so eminently concerned with the work of the senses, with sensory meaning, interpretation, and data processing, it is fitting that it should also supply the implicit theory of the senses for which *Ulysses* itself serves as an index and instrument. This theory comes in two installments: first by way of Stephen's digression on perception in "Proteus"; second by way of Bloom's meditation on the senses in the luncheon episode. In "Proteus," the reader finds Stephen walking by himself along the Sandymount strand, and the chapter stages a representation of a sensuous world whose dimensions are perceived in either visual or aural terms. Designated a male monologue by Joyce, the episode opens with Stephen's idiosyncratic yet exacting ruminations on perception. Stephen reflects on the sense of sight and that of sound, one at a time. Thus the two previous chapters' modes of rendering the operations of Stephen's sensory apparatus are both justified and amplified, and this is why "Proteus" represents the climax of his introduction into *Ulysses*. A practical experiment in apperception, the opening of the episode revolves around what Stephen calls the "ineluctable modality of the visible," then proceeds to that of the audible. It thus stages the question of how sight and hearing mediate his knowledge and experience of the physical world, that is, how they read sense data. Joyce, incidentally, designated the "art" of this chapter as "philology." Of the two senses, sight enjoys primacy, and sight also opens Stephen's meditation:

Ineluctable modality of the visible: at least that if no more, thought through my eyes. Signatures of all things I am here to read, seaspawn and seawrack, the nearing tide, that rusty boot. Snotgreen, bluesilver, rust: coloured signs. Limits of the diaphane. But he adds: in bodies. Then he was aware of them bodies before of them coloured. How? By knocking his sconce against them, sure. Go easy. Bald he was and a millionaire, *maestro di color che sanno*. Limit of the diaphane in. Why in?

Diaphane, adiaphane. If you can put your five fingers through it it is a gate, if not a door. Shut your eyes and see. (*U* 3.1–9)

Stephen's reflection is in part a pyrotechnical dialogue with philosophers of perception such as Aristotle (the visible is visible only through color), Berkeley (the visible is a sign language, written by the Author of nature who thereby manifests his spiritual presence) and Jacob Böhme (the theologian who wrote *The Signature of All Things*).[46] Over and above its philosophically inflected ruminations, however, Stephen's monologue is also a discourse on method. In effect, it is a discourse on Joycean method: an aesthetic metacommentary that affirms the linguistic and perceptual terms mediating both Stephen's and Bloom's experience of the sensuous world.

Having reflected upon sight, Stephen feels compelled to theorize and practice hearing, that is, to "shut [his] eyes and see." Hearing is thus rewritten in terms of sight, an imagery that suggests the preeminence of the language of the eye. In effect, sight is so powerful that eyes have to be closed in order for the world to be heard:

Stephen closed his eyes to hear his boots crush crackling wrack and shells. You are walking through it howsomever. I am, a stride at a time. A very short space of time through very short times of space. Five, six: the *Nacheinander*. Exactly: and that is the ineluctable modality of the audible. Open your eyes. No. Jesus! If I fell over a cliff that beetles o'er his base, fell through the *Nebeneinander* ineluctably! I am getting on nicely in the dark. My ash sword hangs at my side. Tap with it: they do. My two feet in his boots are at the ends of his legs, *nebeneinander*. Sounds solid: made by the mallet of Los *demiurgos*. Am I walking into eternity along Sandymount strand? Crush, crack, crick, crick. Wild sea money. Dominie Deasy kens them a'. [. . .]

Open your eyes now. I will. One moment. Has all vanished since? If I open and am for ever in the black adiaphane. *Basta!* I will see if I can see.

See now. There all the time without you: and ever shall be, world without end. (*U* 3.10–28)

Stephen's eyes, not his ears, ensure him that the world is there in all its inert actuality, vigorously resisting all attempts at solipsism. Yet at the same time the world requires advanced visual reading skills, for it appears before him as a universe of enigmatic signatures. Thus posited as a privileged means of knowledge, vision provides an epistemological model; yet it is far removed from the Cartesian model of transcendental vision. Indeed, what complicates matters is that perception, in Joyce as in most other modernist writers, artists, and practitioners, is fundamentally located in the physiological infrastructure of the individual subject, that is, in the subject's irreducible corporeality. What style, then, could match this complex epistemology?

Precious little happens in this philological episode. Stephen notices a lone boot. He watches how two midwives cross his path. He glimpses a man and a woman and their dog. He studies the dog as it runs about on the beach. He listens to the rolling sea. He observes the waves. These perceptual reports are interspersed with erratic thoughts and wandering memories. Meanwhile he scribbles down passing fancies, impressions, and ideas in a notebook, as befits a writer-to-be. Trivial incidents, to be sure, but Stephen's emphatically poetic sensibility fleshes them out, transforming them into events. The episode itself is the evidence. In attempting to reproduce the acoustics of the rolling sea, for example, Stephen produces a sonic poetry, a poetry of assonances and alliterations—all for the benefit of the reader's inner ear: "In cups of rocks it slops: flop, slop, slap: bounded in barrels. And, spent, its speech ceases. It flows purling, widely flowing, floating foampool, flower unfurling" (U 3.458–60).

The sight of the rolling sea, on the other hand, is transformed into a tableau. Projecting a fin-de-siècle image of female passion and dangerous lure onto the appearance of the sea, Stephen once more calls upon the eye to reread and rewrite the seen, turning the sea into a Jugendstil-inspired vision of women: "Under the upswelling tide he saw the writhing weeds lift languidly and sway reluctant arms, hising up their petticoats, in whispering water swaying and upturning coy silver fronds" (U 3.461–63).

While observing how the waves roll to and fro, he also alludes to photographic vision, its freezing of time and of movement, as though his eyes operated like a camera: "Ah, see now! Falls back suddenly, frozen in stereoscope. Click does the trick" (U 3.419–20). It falls to the eye and the ear to carry out such an aesthetic renovation project. The entire episode, not to say the entire novel, bears witness to such aesthetic labor.

Thus turning into a giant prose poem, "Proteus" ultimately celebrates the modernist writer's power and privilege to radically rewrite the experience of the real, to make it new. The abandoned boot, after all, is "rusty," the waves "whitemaned," the sand "shellcocoacoloured." The dog's sonic behavior is similarly rewritten: "The dog's bark ran towards him, stopped, ran back" (U 3.310). Once more a sensory impression is translated into visuality; and this image, sparked by a characteristically Joycean will-to-style, testifies yet again to the preeminence of vision and visuality in Joyce.

Bloom's Theory of the Senses

Stephen's monologue on sight and hearing finds a structural parallel in Bloom's meditation on blindness in the "Lestrygonians" episode, a meditation that turns into a Bloomian theory of the senses. The organ to which the episode corresponds is the esophagus, the major narrative event being Bloom's luncheon. The ad canvasser walks from establishment to establishment, reflecting on food, ads, and the city, until he finally settles for Davy Byrne's pub.

At this point in the narrative, Bloom has spent half a day in the city. He has traversed sectors of Dublin several times, either on foot or traveling in vehicles. Continuously compelled to read the visual environment, its "signatures" and opaque messages, he has had to decode those lifted hats, process that ad on the wall, decipher the rituals of the holy communion, and so on. Now Bloom, having just left his luncheon place, spots a blind young man about to cross the busy street. The "blind stripling" accepts Bloom's offer to see him across. "[Bloom] touched the thin elbow gently: then took the limp seeing hand to guide it forward" (*U* 8.1090–91).

The "limp seeing hand" is more than a synaesthetic metaphor; it triggers a series of scattered speculations on the implications of blindness, among them synaesthesia. Soon enough the ad canvasser finds himself pondering matters of epistemology. When the blind man has crossed the street, he says, "Thanks, sir". Bloom infers that the young man's ear compensates for the eye, that the ear ensures the positivity of knowledge that vision normally warrants in matters of sexual identity: "Knows I'm a man. Voice" (*U* 8.1102). This epistemological speculation is fleshed out when a baffled Bloom, having told the blind man to take a left and bidden him farewell, marvels at his sensory competence as he feels his way with the cane: "How on earth did he know that van was there? Must have felt it. See things in their forehead perhaps: kind of sense of volume. Weight or size of it, something blacker than the dark. Wonder would he feel it if something was removed. Feel a gap. Queer idea of Dublin he must have, tapping his way round by the stones" (*U* 8.1107–11).

Soon enough the young man is absorbed by the streets, yet Bloom continues to brood on the implications of blindness, mulling over the question of how the sensuous world appears to the blind. The result is a virtual inventory of the remaining senses, from smell to taste to touch:

Look at all the things they can learn to do. Read with their fingers. Tune pianos. Or are we surprised they have any brains. Why we think a deformed person or a hunchback clever if he says something we might say. Of course the other senses are more. Embroider. Plait baskets. People ought to help. Workbasket I could buy for Molly's birthday. Hates sewing. Might take an objection. Dark men they call them.

Sense of smell must be stronger too. Smells on all sides, bunched together. Each street different smell. Each person too. Then the spring, the summer: smells. Tastes? They say you can't taste wines with your eyes shut or a cold in the head. Also smoke in the dark they say get no pleasure.

And with a woman, for instance. More shameless not seeing. That girl passing the Stewart institution, head in the air. Look at me. I have them all on. Must be strange not to see her. Kind of a form in his mind's eye. The voice, temperatures: when he touches her with his fingers must almost see the lines, the curves. His hands on her hair, for instance. Say it was black, for instance. Good. We call it black. Then passing over her white skin. Different feel perhaps. Feeling of white. (*U* 8.1115–31)

Given that Bloom moves through an urban space which, throughout the novel, is represented as overflowing with retinal stimuli, and that Bloom himself displays such a striking penchant for visual activity, especially of a voyeuristic and libidinally inflected kind, it is significant that Joyce has Bloom produce digressions on blindness. In fact, the blind man is a structural antidote to Bloom; the former's mode of operating in a nonvisual world turns Bloom's sensory *modus operandi* inside out. Thus the stripling, as Keith Cohen suggests, is the "precise opposite of Bloom," who "watches all but can hear only in bits and pieces."[47]

As Bloom speculates on the repercussions of blindness, he effectively rediscovers sight: a different experience of Dublin, a different form of sexual encounter, a different kind of cigarette flavor, and so on. In other words, Bloom's meditation on blindness is also a meditation on sight. And although Bloom's musings waver between a fear of deprivation (a diminished sense of taste, for example) and a fascination with reinforced tactile sensations (without sight, sex is supposedly better and more daring), his long reflection on the blind man ends in a rhetoric of anxiety. Thus, after he has practised a little blindness for himself ("he slid his hand between his waistcoat and trousers and, pulling aside his shirt gently, felt a slack fold of his belly. But I know it's whitey yellow. Want to try in the dark to see" [*U* 8.1140–42]), Bloom conjures up dread: "Poor fellow! Quite a boy. Terrible. Really terrible. What dreams would he have, not seeing? Life a dream for him. Where is the justice being born that way? All those women and children excursion beanfeast burned and drowned in New York. Holocaust. Karma they call that transmigration for sins you did in a past life the reincarnation met him pike hoses. Dear, dear, dear. Pity, of course: but somehow you can't cotton on to them someway" (*U* 8.1144–50).

By contemplating lack of vision as a curse visited on the living, as fate and disaster, Bloom's excursus on the consequences of blindness ultimately testifies to the sublime importance of sight—in Bloom's own world as well as in the Joycean universe at large.

Ulysses, as we have seen, is a self-referential epic of the human body. Joyce's modernist tale not only plots how the senses set to work in a visually dominated urban environment; it also provides comment on the two sensory modalities—sight and hearing—that are selected for the utopian task of reconfiguring the comprehension of the world in their own image.

The Female Gaze

Yet although the separation of the eye and the ear, including their concomitant autonomization, marks the entire narrative, it is far from globally distributed among the characters in *Ulysses*. In effect, there is a sexual division of labor at work. The male characters, notably Stephen and Bloom, perceive the world "cinematically" (optical space, as Bazin puts it, is centrifugal; vision is framed; the seen is divided into parts and autonomized; and so on).[48] Meanwhile, the major female protagonists, Molly Bloom and Gerty MacDowell, tend to perceive the world in what seems a more "organic" manner—or, what amounts to the same thing, the not-yet-modernist manner. In Joyce, therefore, visuality—the gaze, the look, the self-sufficient eye—is not only a locus of gender and gendering, but also a locus of gendering forms of aesthetic activity. That is to say, the gaze not only operates in a gendered visual economy which determines who is to look and who is to be looked at, but also in a system which codes the visual aestheticization of the real as masculine, and organic modes of looking as feminine.

This becomes especially evident in "Nausicaa," the episode whose subject matter is Bloom's masturbation. There, Bloom derives sexual pleasure from looking at Gerty's legs; and she, in turn, derives pleasure from being looked at.[49] Set in a melodramatic register, the episode mocks not only the melodramatic forms of literature which are Gerty's favorite reading, but also a certain mode of visual activity. Bloom is ridiculed, to be sure, but above all Gerty is. For Gerty's gaze is represented as doing precisely what Bloom's eyes (and Stephen's) never do: her gaze carries a hermeneutics and projects knowledge onto the object seen, allowing her to draw certain conclusions concerning Bloom's mood: "she ventured a look at him and the face that met her gaze there in the twilight, wan and strangely drawn, seemed to her the saddest she had ever seen" (*U* 13.368–70). Indeed, unlike Bloom (and Stephen), Gerty looks primarily in order to derive or secure knowledge from what she sees, not for the sake of looking or aestheticizing:

> And while she gazed her heart went pitapat. Yes, it was her he was looking at, and there was meaning in his look. His eyes burned into her as though they would search her through and through, read her very soul. Wonderful eyes they were, superbly expressive, but could you trust them? People were so queer. She could see at once by his dark eyes and his pale intellectual face that he was a foreigner [. . .]. He was in deep mourning, she could see that, and the story of a haunting sorrow was written on his face. (*U* 13.411–22)

Gerty's visual world is one of recognition, habit, bias, prejudice, received knowledge, and banal assumptions, all illuminated by her attempts to read Bloom's gaze and external appearance. For her, clearly, the separation of the observed and the known has not yet happened; and for this reason, her perspective emerges as a striking counterpoint to those of Joyce's male heroes. In effect, Gerty MacDowell opens a window onto the parameters of Joyce's aesthetics of visuality. It would take another study to spell out the consequences of this gendered economy of visual perception for the aesthetic program implicit in *Ulysses*.[50] It is clear, however, that the look operates in a signifying system that associates the aestheticization of the real with a masculine point of view. Perhaps needless to say, the masculine perspective is charged with a positive value. It is Joyce's male protagonists who carry out the task of renovating the world, ineluctably remaking it in the name of the visible and the audible, thus attempting to undo the kind of habituated perception for which Gerty MacDowell and her not-yet-modernist sisters serve as an emblem.

II. The General

In *Ulysses*, as we have seen, the description of how the male protagonists process the sensuous world is routed through the notion of the embodied individual. Activities of seeing and hearing emerge as rigorously corporeal activities, primarily aimed at perceiving for perception's sake. Separated from the other senses, the body at large and, often enough, the cognitive faculties to which they are nevertheless related, the sense of sight and that of hearing tend to operate autonomously. The eye even becomes a narrative character in itself, endowed with an agency all its own.

Joyce's implied narrator not only describes modes of perception, making perceptual activities a vital part of the novel's subject matter; the narrator also describes objects of perception in ways that imply certain modes of perception. Indeed, from the ways in which objects of perception are rendered, one may infer highly specific modes of perception. The solitary finger, the independent tongue, the automatic teaspoon, the thundering sound of voices and boots: such modes of representation presuppose autonomization, leveling, and framing. As I have suggested, these recurrent stylistic features are enabled by those matrices of perception that accompany technologies for reproducing sense data such as phonography and cinematography.

Joyce's style, then, is a diagram of the historical processes that the novels of Mann and Proust reflect on: the dissociation of perception and knowing, the differentiation of the senses and, by implication, the marginalization of the mandates of the human senses in an age when inscription devices and vision machines increasingly claim superiority if positive knowledge is to be gained and ascertained. If, then, the act and means of sense perception have become problems unto themselves, *Ulysses* registers and amplifies that process all at once. Joyce's modernist tale thus reflects and reinscribes a social history of the sensorium. As I will discuss, this historical process elicits Joyce's aesthetic imperative: to make the reader see for the sake of seeing and hear for the sake of hearing.

So far we have been dwelling on the level of the sentence, exploring how the ever-accumulating mass of details suspends the potential whole. It now remains to discuss the level of the larger formal structure. For where the level of the sentence tends toward differentiation and autonomization, the overall form of *Ulysses*—the organic ideal—dreams of synthesis and synaesthesia, projecting a solution in the form of the *Gesamtkunstwerk*, the total work of art.

Joyce conceives his literary artifact in a historical situation that bears witness to a radically destabilized economy of media and modes of cultural production. Less than a century after Hegel had formulated his philosophical aesthetics, Guillaume Apollinare in 1917 delivered a prophecy that throws Hegel's idealist aesthetics into relief. Where Hegel, in his *Ästhetik*, argued that the eye and the ear were "theoretical" and hence "ideal" aesthetic organs, Apollinaire, himself a poet and critic, anticipated an age where technological modes of storing and reproducing the visual and aural real had achieved hegemony. And where Hegel held that spoken poetry represents the most ideal art form because of its unique, sensuous presence in time and space, Apollinaire forecast an age where both spoken poetry and its supplement—the printed book— would have been marginalized, not to say extinguished: "[The book] is declining. Within one or two centuries it will die. It has found its successor, its only possible successor, in the phonographic disc and the cinematographic film. One will no longer need to learn to read and write."[51] Apollinaire does not reflect upon Spirit, to be sure, but it goes without saying that Hegel's notion of genuine art—that it be the sensuous objectivation of *Geist* and, by the same token, a unique spatiotemporal presence—will long since have been eclipsed.

The modernist novels, however, rise to the challenge. Defying the emerging technological means of reproducing the real, they swell into literary counterparts of the Wagnerian *Gesamtkunstwerk* and its ideal of the ultimate synthetic art form.[52] Joyce's *Ulysses*, Proust's *Remembrance*, Mann's *Magic Mountain*, Musil's *Man without Qualities*, Broch's *Sleepwalkers*, these novels all incorporate within themselves representations of other modes of cultural production, including competitors such as the newspaper, cinematography, and recorded music, attempting to overarch their specific idioms and modes of expression, at the same time dreaming of a total sensibility that will abolish the increasing specialization of the arts and modes of communication and, at length, reinstate the legitimacy of the art of the novel.

Joyce and the Total Work of Art

Early in *Ulysses*, the reader finds Bloom seated on the toilet in the privy in his garden, meanwhile reading a newspaper. Detailed reports on the progress of Bloom's bowel movements are interfoliated with equally detailed reports on his musings apropos of the contents of the paper. Mulling over the featured prize story, "Matcham's Masterstroke," Bloom thinks to himself: "Print anything now. Silly season" (*U* 4.512). Judging the story quite bad, he plays with the idea of writing one himself. That Joyce would have Bloom deliver a critique of newspaper editing in a scatological context is no coincidence. And if the newspaper world looms large in *Ulysses*—Bloom, recall, makes his living as an ad salesman for the *Free-*

man's Journal—it is still merely a local story within Joyce's global version whose modes of representation ultimately declare themselves superior to such modes of storytelling as journalism.[53]

But Bloom's toilet comment is also a gloss on the aesthetic pursuits and methods of the novel itself. "Print anything now. Silly season." First of all, this is a local metacommentary—on Joyce's decision to zero in on Bloom's defecation. Moreover, it is a global metacommentary—on the deeper significance of Joyce's totalizing desire, on the inexhaustible ambition to record the lived minutiae of a day in the life of a group of Dubliners. In this sense, Joyce's inventory comes across as a chronicle, a testimony. And if all writing is a form of mnemotechnics, *Ulysses* even turns the question of memory into a theme, at the same time providing a running commentary on mnemotechnics.[54]

The thematics of memory makes itself felt especially in the second half of the novel, notably in "Ithaca," where the word "mnemotechnic" is key. There, a plethora of mnemotechnological phenomena are evoked, all the way from hieroglyphics and musical scores to photography. Yet, one of the more remarkable moments has already occurred in "Hades," when, after Paddy Dignam's burial, Bloom lapses into a fantasy that revolves around the idea of memory and the machine. When Dignam's coffin has been sunk into the ground, the gravediggers start filling the grave with "heavy clods of clay," and Bloom, watching their work, cannot help thinking that Dignam might be buried alive: "Whew! By jingo, that would be awful! No, no: he is dead, of course. Of course he is dead. Monday he died. They ought to have some law to pierce the heart and make sure or an electric clock or a telephone in the coffin and some kind of a canvas airhole. [. . .] The clay fell softer. Begin to be forgotten. Out of sight, out of mind" (*U* 6.866–72).

The dead start to sink into oblivion, Bloom thinks to himself, as soon as the grave has been filled; and if the dead are not really dead, a telephone might come in handy. This fantasy, associating death and telecommunications, is all the more remarkable for a narrative like *Ulysses* in which, after all, activities such as talking on the phone, riding the tram, deciphering advertisements, reading newspapers, and typewriting are represented in their lived immediacy, as fully naturalized everyday practices. In *Ulysses*, a telephone call is no more strange, but also no more familiar, than, say, kidney-shopping, watching someone climb the stairs, looking for a coin in one's pocket, scratching one's stomach, or simply breathing. This is how Miss Dunne's telephone conversation in the newsroom is reported:

—Hello. Yes, sir. No, sir. Yes, sir. I'll ring them up after five. Only those two, sir, for Belfast and Liverpool. All right, sir. Then I can go after six if you're not back. A quarter after. Yes, sir. Twentyseven and six. I'll tell him. Yes: one, seven, six.

She scribbled three figures on an envelope.

—Mr Boylan! Hello! That gentleman from *Sport* was in looking for you. Mr Lenehan, yes. He said he'll be in the Ormond at four. No, sir. Yes, sir. I'll ring them up after five. (*U* 10.389–96)

True to his mode of impersonal, non-omniscient narration, Joyce's implicit narrator reports the one side of the conversation, and that alone, as though his task is to record what he has heard, not what he knows is there. As a result, the possible meaning of the phone conversation appears about as fragmentary to the reader as it would to someone who merely happened to pass by Miss Dunne's desk. The passage simulates phonographic modes of reproducing acoustic phenomena, for the narrator's ear operates as indiscriminately as would a phonographic recording device. It is as though the scene has been recorded and stored for future editing.

Since the new technologies that appear in *Ulysses* are generally integrated into the fabric of everyday life, Bloom's fantasy about death and telephony would seem to deviate from this pattern. On the other hand, Bloom's rumination shares in a historically mediated imaginary, an imaginary in which death is part of a cluster of ideas that gather around the image of technology. As Avital Ronell suggests, the telephone "has always been inhabited by the rhetoric of the departed."[55] The telephone episode in Proust's *Guermantes Way* offers a rich case study. There, the anonymous telephonists are projected as pagan goddesses, as "priestesses of the invisible," thanks to whom our friends and loved ones are resurrected from a world of shadows; and, as we have seen, the narrator's phone conversation with his grandmother forebodes that eternal separation called death.

An even more exemplary case is Walter Rathenau's 1898 short story "Die Resurrection Co."[56] Set in a place called Necropolis, it tells the story of the most technologically advanced funeral ceremony in the United States. In fact, it is so speedy that the undertaker, in order to prevent his clients from being buried alive, decides to have electric wiring, telephones, and bells installed in each grave. Late one evening No. 169, a gentleman nine months dead, rings the bell. The case is investigated, yet no error is found. Two weeks later No. 289 calls, a Miss Simms, also several months dead, asking to be connected to No. 197 . . .

Rathenau's imagery, like Proust's, is inscribed in a collective fantasy formation—the technological uncanny—which turns on the idea of the double, the copy, and the automaton. Bloom's musings on the telephone at Dignam's burial also belong here. In the midst of these ruminations on death and underground communication, however, Bloom is interrupted by a journalist, Hynes. There to report on the funeral, Hynes scribbles down facts in a notebook, asking Bloom how to spell his name and other questions. From a formal point of view, however, the fact that Hynes crosses Bloom's path is no interruption. Indeed, their encounter amplifies the emerging thematization of memory and mnemonics, and so it is that the idea of the newspaper report as a form of testimony joins the thematic cluster that includes death and technology. The sixteenth episode of *Ulysses*, "Eumaeus," relates how the ad canvasser, in the early morning hours, reads Hynes's report as printed in the paper, and this incident, as I will discuss, offers a significant moment in *Ulysses*.

Having taken down the necessary information, Hynes vanishes with his mostly faulty notes, and Bloom walks away by himself, pondering mortality, heaven, and money as he crosses the fields. Bloom continues his meditation on the dead, reflecting upon their potential, indeed spectral, afterlife in the world of the living:

How many! All these here once walked round Dublin. Faithful departed. As you are now so once were we.

Besides how could you remember everybody? Eyes, walk, voice. Well, the voice, yes: gramophone. Have a gramophone in every grave or keep it in the house. After dinner on a Sunday. Put on poor old greatgrandfather. Kraahraark! Hellohellohello amawfullyglad kraark awfullygladaseeagain hellohello amawf krpthsth. Remind you of the voice like the photograph reminds you of the face. Otherwise you couldn't remember the face after fifteen years, say. (*U* 6.960–68)

In juxtaposing the gramophone and the telephone, Bloom articulates a germinal theory of technology.[57] The gramophone, as Bloom's double analogy suggests, is to the voice as the photograph is to the face—both being "reminders" that prevent the work of forgetting, potential mediators between the present and the past, the living and the dead. Of course, Bloom's mundane fantasy about the destiny of the dead is a mockery, especially of the gramophone. Located now in the silent grave, now in the bosom of Sunday domesticity, the gramophone emits crackling echoes of a past that penetrates the present. Sonorous traces of the now dead come to inhabit the temporality of the moment. The mechanical reproduction device thus both doubles the dead and others the past; in a word, the gramophone spectralizes the present.

In this Bloomian fantasy, then, the themes of death, memory, and mechanical storing devices are grouped together, testifying to the legacy of the technological uncanny. But the passage also raises another issue—the question of memory and, therefore, of experience and history. Stated simply, remembrance is presented as a problem. And here as elsewhere in *Ulysses,* the problematization of memory has to be considered in light of the fact that in Joyce, phonography and gramophony are prior to the notion of voice, just as photography and cinema precede the notion of face.

In Proust, too, memory is a problem. If the advent of technologies for storing and reproducing sense data reopens the question of human memory, helps problematize its nature and status, and contributes to the differentiation of its very idea, then it is easy to understand the large extent to which history fuels the great theme of recollection and retrospection in *Remembrance.* For Proust, the notion of *mémoire involontaire* offered a solution to the question of genuine memory, as did the inextricably intertwined idea of chance sensory experience. In *Remembrance,* it falls to the five senses, and to them alone, to warrant experiential authenticity and provide immediate access to the past. The burden of Proust's novel is, among other things, to articulate a phenomenology of pure memory, a memory that is immediate in the sense that it is mediated only by the individual body and its infrastructure. But Proust's solution cannot be Joyce's, not in a work like *Ulysses* that so resolutely and obsessively turns the labor of the senses into a subject matter in its own right and which thus both registers and articulates their reification.[58] To move from Proust to Joyce, then, is to bear witness to a dialectical leap. Simply put, the forms of historical change that Proust's work reflects upon make themselves felt in the formal aspects of Joyce's novel. The labor of the senses meanwhile has been transformed into a new kind of narrative content, for in Joyce, as I have argued, the representation of the work of the senses becomes an end in itself.

Now Joyce's concern is not primarily with the past, nor with time as such, but rather with the ways in which experience is derived, the continuous making and unmaking of immediate experience in a historical situation in which technological devices are capable of storing and reproducing sensations, at the same time articulating new perceptual realms. To be more specific, Joyce's concern is with the *naming* of immediate experience, hence also with the writing of experience.

The newspaper, for example, names experience. About one o'clock in the morning, Stephen and Bloom take to a cabman's shelter, where Bloom happens upon the late edition of the *Telegraph*. His "eyes went aimlessly over the respective captions which came under his special province," Joyce writes. A variety of current topics, large and small, high and low, are listed as they pass before Bloom's critical gaze (*U* 16.1236–37). The diversity and randomness of topics, including their arbitrary constellations, reflect the utter heterogeneity of materials to be found in *Ulysses* as well. What is more, some of the news items are recognizable parts of the thematic texture of the novel itself, such as foot-and-mouth disease, the New York disaster, and the funeral.

The passage details how Bloom reads the news item about the deceased Paddy Dignam ("a most popular and genial personality in city life") and his funeral service ("many friends of the deceased were present," among them "L. Boom"). Quoted line by line, the citation of the report is interspersed with Bloom's reflections ("Hynes put it in of course"). This quotation thus reiterates an event that Joyce's narrator has rendered already, and by including Bloom's reading of the daily paper, *Ulysses* both cites itself and provides a summary of some of the events and incidents that make up the narrative, now framed by the editorial principles that inform newspaper writing. At length, the quotation of Hynes's report emerges as a textual fragment in a labyrinthine hall of mirrors, reflecting obliquely the mode of writing in which Dignam's funeral was first presented, as well as Joyce's global narrative, that literary *Gesamtkunstwerk* which, weaving its way through style after style, technique after technique, art after art, ultimately rises above them all and so relativizes them. In this way, Joyce's *Ulysses*, like Mann's *Magic Mountain* and Proust's *Remembrance*, turns itself into an all-encompassing literary artifact, an ambitious literary encyclopedia that declares the received idea of the novel obsolete and in any event inadequate to the task of capturing modern everyday life.

Significantly, in the boundaries of *Ulysses* the entire history of Western literature from Homer and Goethe to Maeterlinck echoes, including a vast variety of literary genres and subgenres such as drama, poetry, limericks, and songs.[59] Also, *Ulysses* systematically attempts to exhaust the languages, styles, and idioms to be found in newspapers, advertising, poems, newspapers, letters, telephone conversations, telegrams, protocols, budgets, interrogations, letters, postcards, and so on. By the same token, Joyce's epic is an inventory of colloquial as well as scientific modes of expression, bureaucratic as well as Shakespearean speech. As Colin MacCabe puts it, "*Ulysses* is a voyage through meaning: a voyage through all the discourses available in English in 1904."[60]

Joyce's encyclopedic work of literature, however, engages in mimesis in another sense. It simulates cinematographic modes of reproducing the visual, just as it simulates phonographic modes of reproducing the aural. In other words, by incorporating within itself cinematic modalities of the visible and phonographic modalities of the audible, *Ulysses* emulates crucial features of those representational techniques inherent in the two new modes of cultural production that, according to Apollinaire, could lead to the extinction of the book. Indeed, from drama to cinematography, from poetry to phonography, Joyce's epic excels in the reproduction of other modes of mimesis, ultimately aiming at creating a radically new literary artifact that incorporates within itself the various modes of reproduction with which it competes.

In bringing various artforms into contiguity with one another, *Ulysses* attempts to transcend the increasing differentiation of modes of cultural production in the age of mechanical reproducibility, notably their increasingly specialized ways of addressing the senses. As a *Gesamtkunstwerk,* Joyce's novel is synaesthetic in more ways than one. Not only does it bring different forms of art, high and low, old and new, into proximity; it also attempts to call forth something like a synaesthetic experience. Within the abstract world of the printed word, *Ulysses* endeavors to cater to the eye and the ear, continuously supplying the quasi-ideal senses of sight and hearing with sublime pleasures: visuals to be seen for the sake of seeing (Molly's spoon, Bloom's fingers, Brayden's umbrella), and sonics to be heard for the sake of hearing (voices, onomatopoeia, assonances, alliterations, tapping canes, dullthudding barrels, and so on).

By incorporating newspaper headlines and musical scores, *Ulysses* even attempts to offer the textual equivalent of tactility. Granted, this is an essentially visual form of texture, an optical pleasure that has to be refracted through the reader's eye to become effective. Indeed, Joycean visuality and aurality have to rely upon the eye—an eye that translates the diverse representations of sight and sound into mental sensations to be seen or heard in the silent interiority of the reader. That is to say, just as Joycean tactility is necessarily rerouted through the sense of sight, so too the immediacy of visuality and aurality require mediation. It is the printed word, and that alone, that carries the burden of utopian synaesthesia in *Ulysses.* The irony is both irreducible and inescapable. In the end, as Susan Stewart reminds us, "the printed word always tends toward abstraction, for it escapes both the necessity of a material referent and the constraints of an immediate context of origin."[61]

With Joyce the art of the novel is taken to a point where its very idea is exploded—Eliot could not imagine Joyce writing another "novel."[62] In fact, not only the art of the novel is at stake, but the notion of art as such. From Proust to Musil, from Mallarmé to Marinetti, from Conrad to Eisenstein, from Vertov to Moholy-Nagy: the romanticist ideal of the synthetic and synaesthetic work of art is revitalized with unprecedented urgency.[63] Art is assigned the task of overcoming what is perceived as the compartmentalization of life praxis, particularly the separation of art and everyday life. Moholy-Nagy, in his Bauhaus treatise *Painting, Photography, Film* (1927), discusses the two paths that had been explored in the attempt to recuperate the function of art and make it an integral part of life praxis: on the one hand, the avant-garde impulse to negate the idea of art, and on the other, the unifying impulse to bring together separate fields of artistic creation under the umbrella of the total work of art, the *Gesamtkunstwerk*. For Moholy-Nagy, neither is a viable solution. The avant-garde is inevitably locked into the bourgeois notion of art as leisure that it is trying to explode; the *Gesamtkunstwerk* is still confined to the traditional domains of art, thus essentially divorced from life praxis. Moholy-Nagy promotes instead the cultivation of a synthetic sensibility that is to express itself, not in the *Gesamtkunstwerk*, but in the *Gesamtwerk*, the total work, a utopian form of cultural production that, as he suggests, seeks to embrace the "wholeness of life."[64]

Mann, Proust, and especially Joyce reveal an affinity with the characteristically modernist sensibility for which Moholy-Nagy's proposal serves as an emblem. Stretching the genre of the novel to the point of bursting, the narratives of Mann, Proust, and Joyce negate the generic rules that went before them, attempting to encompass human experience in its lived totality. Just as *The Magic Mountain* and *Remembrance* are essentially self-contained and self-referential novels, producing within themselves leitmotivs that elicit new layers of signification, so *Ulysses* creates a literary universe that builds its own circuits of meaning. Joyce, however, questions the idea of genre even more emphatically than do Mann and Proust. A fundamentally self-referential textual artifact, *Ulysses* recognizes no alternative other than itself. It refuses to offer itself as an example. In keeping with the attempt to elaborate an aesthetics of immediacy, *Ulysses* seeks to be immanent to itself.

Significantly, *Ulysses* proposes that the notion of individual style is no longer valid. Indeed, once we reach *Ulysses*, style is no longer a question of writing well. Proust, for one, could still insist that a writer should aspire to writing well. In his view, a good writer strives for perfection in matters of style, not by imitating achievements in the past but by inventing, indeed by making it new: a new style, a new inflection, a new language. Promoting an anticlassical ideal, Proust, like so many other writers at the time, declared the time-honored ideal of emulation obsolete. Explaining his notion of the well-written to a friend, Proust evoked and emphasized the idea of originality:

> Every writer is obliged to create his own language, as every violinist is obliged to create his own "tone." [. . .] I don't mean to say that I like original writers who write badly. I prefer—and perhaps it's a weakness—those who write well. But they begin to write well only on condition that they're original, that they create their own language. Correctness, perfection of style do exist, but on the other side of originality, after having gone through all the faults, not this side. [. . .] The only way to defend language is to attack it.[65]

Once we reach Joyce's *Ulysses,* the dialectic of the new and the old, the original and the conventional, has been exhausted. It has reached a trigger point that sparks a qualitative change in terms of style, for even if style is at the center of Joyce's titanic enterprise—it is, after all, its primary subject matter—the idea of diversity has been substituted for that of originality. Instead of perfecting a new and original style, Joyce delivers an encyclopedia of recognizable styles. The classical ideal of emulation thus returns, but with a difference: Joyce's modernist artifact vigorously seeks to transcend the domain of literature, even the domain of art as such, attempting to capture the experience of everyday life in its lived immediacy and, in order to do so, grafts upon itself those modes of representing the real provided by those technologies that threaten to marginalize the privileges of the written word: phonography and cinematography.

It is true that Joyce and Proust both endeavor to represent the immediacy of human experience, but Proust, unlike Joyce, ultimately also seeks to reflect upon its conditions and enabling constraints. To be sure, Proust's contemplations on the experience of the advent of technological phenomena are posited as essentially particular experiences—the narrator's riding in a motorcar, seeing an airplane, and speaking on the telephone are all framed by a rhetoric of the-first-time ("her voice itself I was hearing this afternoon for the first time"). Yet the very notion that these events signify a first time make them representative of an implied historical narrative; indeed, among all the stories that make up *Remembrance,* there is also the story of the historicity of forms of experience.

Joyce, however, refuses to mediate. This accounts for one of the vital differences between Joyce on the one hand and Proust, Mann, and most other modernist writers on the other. Joyce seeks the general in order to particularize it, all in an effort to make the reader *see* and *hear*, as though for the first time, and his prime means is language. Yet language, as Adorno once suggested, necessarily mediates the particular through the general (*das Allgemeine*). The constitutive paradox from which much of modernist literature derives its vitality, then, is that language is "hostile to the particular and nevertheless seeks its rescue."[66] Joyce, incidentally, was once asked whether there were not enough words in the English language for him. "Yes," he retorted, "there are enough, but they aren't the right ones."[67]

In renouncing the idea of stylistic originality in favor of systematic emulation and relentless diversity, Joyce attempted to suspend the generalizing tendency of language—all in an effort to carve out a space for the possibility of representing immediate experience in a historical age in which the category of experience itself had become a problem. The fact that Joyce developed stylistic counterparts to the matrices of perception accompanying those technologies that contributed to the abstraction of sensory experience accounts for the historical originality of *Ulysses,* that eminently modernist epic of the human body and its adventures in the second machine age.

Coda

The Legibility of the Modern World

Like a host of European writers born in the late nineteenth century, Mann and Proust chronicle how the new collides with the old. Staging first encounters with various technologies, they contemplate how the new machines and their environments alter the ways of the world. Traditional modes of experience flare up one last time, visible for yet a few moments until the new technologies have been domesticated by the cognitive and perceptual habits they simultaneously help to transform. Those older worlds of habit, now subject to transfiguration, are henceforth experienced as having been organic. Technology thus enters the novels of Proust and Mann as an emblem of historical rupture. In fact, in modernism at large the image of the machine is so often grouped together with images of mortality, finitude, and spectrality that the constellation appears as a standing figuration. Consider, for example, the allegorical killing machine in Franz Kafka's short story "In the Penal Colony" (1919), or the furious attack on cinema as the Hades of modern man in Joseph Roth's *Antichrist* (1934), or the deadly automobile chase in Herman Hesse's *Steppenwolf* (1927)—"I saw at once that it was the long-prepared, long-awaited and long-feared war between men and machines, now at last broken out," as Hesse writes.[1] For all its existential overtones, the recurrent machine-death figure has to be accounted for as a way of managing and making sense of those historical processes called modernization.

Technology helps change not only the world but also the perception of that world. This is partly why the image of the machine enters modernism together with problems of intelligibility. In *The Magic Mountain,* the hero's first experience of X-ray photography is a focal point, a narrative event that ultimately raises questions about identity and subjectivity, reading and interpretation, including the status and future of culture. It is no coincidence, then, that Hans Castorp's educational sojourn at the sanatorium should also comprise encounters with two other late-nineteenth-century inventions, two radically new modes of cultural production: cinematography, that reproducible mimesis of the visual world as it moves through time; and phonography, that reproducible mimesis of the audible world as it evolves in time. In *The Magic Mountain,* technology bursts open spaces of interpretation and calls for new modes of managing and producing meaning. At the same time, it emerges as a reminder of death, that is, as an end of all meaning and hence as an index of the vanishing of an older, specifically German *Kultur,* including its inherent modes of perception.

What Adorno once suggested apropos of Proust is therefore partly true of Mann as well. Proust's modernness, he argued, derives from the fact that Proust, born in 1871, looked at the world with the eyes of someone thirty or fifty years his junior and that he, for this reason, represents both a "new stage in the novel form" and a "new mode of experience."[2] Like Mann, Proust dwells on first encounters with recently introduced technological phenomena, particularly the ways in which they call forth new modes of experience. In *Remembrance of Things Past,* the narrator chronicles the sublime appearance of an airplane, just as he explores the fact that his sense of sight has not yet accustomed itself to speedy vehicles and the delicious metamorphosis of the surrounding landscape. The narrator also details how, in a hotel in Balbec, he experiences his first ride in a lift. Finally, he reflects on his first telephone conversation with his grandmother and her disembodied voice at the other end. In all these situations, the new is caught with the old as though in a freeze frame, soon to be broken open by the passing of time. Habit has not yet domesticated sensory experience. For yet a little while, experience is novel, fresh, and unexpected, hence open to aesthetic exploration. "It took Proust to make the nineteenth century ripe for memoirs," Benjamin once observed.[3]

Very little of the nineteenth century seems to remain in James Joyce's *Ulysses.* T. S. Eliot even suggested that Joyce's novel would be a landmark precisely because it "destroyed the whole of the nineteenth century."[4] To be sure, the occasional cow herd makes its way through the Hibernian metropolis; a considerable number of vehicles in the streets are horse-driven; and at dusk the streets are lit by lamplighters. Still, Joyce's Dublin comes across as essentially modern. Telephone conversations are perfectly natural; typewriters are no longer a novelty; gramophones can be heard at night, at least in the red-light districts of Dublin; and advertisements stand for a vital part of the visual life of the city.[5] Last but not least, there is the transit system. For although the motorcar has yet to gain the upper hand in the city inhabited by Leopold Bloom and Stephen Dedalus, another modern technology of speed—the tram—is part and parcel of the experience of everyday life as it appears in *Ulysses*.[6] As Hugh Kenner has proposed, the day that Joyce's novel reflects "would have been impossible a generation earlier, before electric trams were moving people quickly about a large city."[7] The tram links sectors of the city just as it links parts of the novel, thus functioning as connective tissue in Shklovsky's sense, or, to use another somatic metaphor, as a nervous system. Motivating the assemblage of events that make up the narrative, the tram produces the network of associations it simultaneously traces. What is true of the tram also holds for the telephone, the newspaper, and the postal services: they are all modes of communication, all located "in the heart of the Hibernian metropolis," as the opening of the newspaper episode reads.

This means that most of the technological events recorded, thematized, and theorized in Mann and Proust have already occurred in Joyce's urban galaxy. Whereas in Proust the telephone and the automobile trigger reflections upon the historicity of modes of experience, thus modulating one of the great themes of *Remembrance*, in Joyce, by contrast, activities such as talking on the phone or riding in a tram are represented as fully naturalized everyday practices. The dialectic of the new and the old has been exhausted, suspended in favor of a now that continually reinvents itself. This helps explain why Joyce's narrator has withdrawn from the task of commentary, content merely to focalize the action through his protagonists. But the crucial difference lies on a different, though related, level. If Joyce sinks telephone conversations and tram rides—indeed, every conceivable human activity—into the ever-pregivenness of everyday praxis, it is because Joyce, unlike Mann and Proust, pursues the absolute immediacy of experience.

Yet although modern technology appears as an integrated part of human praxis in *Ulysses*, it does not mean that its resonance is less palpable, nor that its effects are less pronounced. Technologies for reproducing sound and vision have left their mark on some of the most characteristic stylistic techniques in *Ulysses*. Indeed, Joyce's style registers, on the level of the sentence, the subterranean effects of those technological events that the novels of Mann and Proust subject to thematization. Joyce's aesthetics comes into being as a solution to a historical problem—how to recover and represent the immediacy of lived experience in an age when modes of experience are continually reified by, among other things, the emergence of technologies for reproducing the visual and audible real. This, then, reveals a historical and thematic continuity between Mann and Proust on the one hand and Joyce on the other.

Meanwhile, in all three novels technology also fuels the imperative to make the phenomenal world new. Indeed, technology emerges as an occasion for launching new aesthetic idioms. Restructuring the prose of the world, it yields opaque signatures that demand to be read, decoded, and often rewritten. No wonder, then, that *The Magic Mountain, Remembrance,* and *Ulysses* should also foreground the thematics of legibility, inscription, and interpretation. Deleuze, we recall, suggests that *Remembrance* is a narrative about the narrator's apprenticeship in the art of reading and producing signs. The Martinville episode, as we have seen, offers a most significant occasion. Deriving from Proust's *Figaro* article on the visual thrills of motoring, the episode signifies how the young narrator discovers a new world of signs and produces his first piece of writing. As the narrator plunges into a new optical domain, Proust enters a space-time vocabulary burst open by technologies of velocity and articulated by cinematographic representations of movement.

The novels of Mann and Joyce may similarly be understood as stories of apprenticeship in the art of reading and producing signs. Here, too, the problem of legibility is clustered together with the image of technology. In *The Magic Mountain,* the protagonist not only learns to read the thermometer and X-ray films; he also becomes an avid reader of scientific books whose systems of knowledge explain the universal order of things. His educational process consists in his having to pass through and identify a series of visually determined signifying systems that redefine criteria of truth and positive knowledge, all the way from the visual hermeneutics of class and the structured medical gaze to radiographic means of representation.

In *Ulysses,* finally, the two male heroes are continuously engaged in the decoding and production of signs as they move through a phenomenal universe of rigorous opacity. Coming before Joyce's protagonists in all its ostensible concretion and particularity, the sensuous world produces a diverse array of signifiers that seek a matching signified—as in the sequence when Bloom, noticing so many hats, considers first their hatness and only then the implied meaning: tributes to the deceased Paddy Dignam. Yet the ways in which Stephen and Bloom negotiate this phenomenal world register the cadences of history; indeed, as I have argued, they must be understood in tandem with the matrices of perception that accompany technologies for recording and reproducing sense data, notably cinematography, phonography, and telephony.

At the same time, the novels of Mann, Proust, and Joyce provide comment on their existence as literary artifacts in a fundamentally altered media economy. This feature in itself motivates a historicizing approach in the effort to understand these prototypically total works of art, including the specific aesthetic sensibilities for which they serve as vehicles. Yet most critical paradigms of literary modernism, based as they often are on the modernists' self-understanding and a general antitechnological bias, have precluded sustained reflection upon the historical conditions and enabling constraints of the modernist enterprise. This is all the more remarkable since the novels of Mann, Proust, and Joyce do more than just comment on the media economy to which they inevitably belong. They also appropriate, even emulate, the modes of perception and representation inherent in cinematography, phonography, and other technologized means of reproducing the real. That is to say, the aesthetic categories in Mann, Proust, and Joyce I have been discussing in the preceding pages are interpenetrated by technological categories. Indeed, my textual analyses not only suggest that there is a constitutive relationship between high-modernist aesthetics and technology, but also that our common understanding of those entities referred to as "aesthetics" and "technology" undergo a mutation as soon as we begin contemplating that insufficiently explored and undertheorized nexus.

The focus on the intersections of technology and modernist aesthetics has also allowed another poorly understood aspect of modernism to emerge into view. As the passage from Mann and Proust to Joyce demonstrates, in the modernist period, the continuity between categories of perceiving and categories of knowing is progressively suspended. This historical process runs parallel to those technological developments that—all the way from, say, the invention of the stethoscope to chronophotography and radiography—help deperceptualize scientific observation in particular and human observation in general. The specifically high-modernist aesthetics of perception I have been exploring thus feeds on a historical irony that is as palpable as it is inevitable: the more abstract the world of observation becomes, the more corporeal is the notion of the perceiver. Indeed, the more the categories of *theoria* become dependent on technological mediation, the more *aisthesis* becomes invested with notions of physiological immanence.

It is no accident that these processes run parallel to another scenario: the ever-closer relationship between the sensuous and the technological.

As the trajectory from Mann and Proust to Joyce suggests, the continuity between perceiving and knowing is dramatically called into question. It also demonstrates that technologically mediated matrices of perception are prone to becoming internalized by the habits of the sensorium. In effect, the move from Mann to Joyce signifies a tendency that stretches from an essentially external relationship between the senses and their technological supplements to an interiorization of technological modes of perceiving. In Mann, to put it schematically, perceptual technologies figure as prosthetic extensions of the human observer, while Proust poses the becoming-prior of technological modes of perception. This development is fully realized in Joyce where, finally, technologically mediated matrices of perception have become at one with *aisthesis* as such. In short, the modernist moment bears witness to a transition from *prosthesis* to *aisthesis*.

More than anything else, it is the eye and the ear that register this complex and overdetermined thematics. This means that the ever-closer relation between the sensuous and the technological traverses the question of aesthetics. Indeed, the ways in which the tasks of the eye and the ear are orchestrated in Mann, Proust, and Joyce imply that modernism is coextensive with a shift from idealist theories of aesthetic gratification to essentially materialist ones. That is to say, modernism is an index of a general gravitation toward a conception of aesthetic experience based in a notion of the immanence of the body, a body inhabited by temporality and therefore also finite.[8]

To appreciate the scale and nature of this shift, it is helpful to consider Hegel's philosophy of art. Throughout the history of aesthetic discourse, sight and hearing have been privileged over taste, smell, and touch. Because they are the least corporeal of the senses, sight and hearing are more readily disposed to abstraction, and this is partly why they have enjoyed such prominence in the history of aesthetics. According to Hegel, sight and hearing are essentially *theoretical* senses. For this reason, they are also *ideal* senses. Taste, smell, and touch, by contrast, are *practical* senses. They involve consumption of the work of art in one way or the other, and thus fail to qualify as aesthetic organs. Only the eye and the ear are capable of respecting the integrity and freedom of the work of art. This is a crucial criterion of aesthetic apprehension, for in Hegel's view, the work of art is an ideal site where spirit and matter intersect. A privileged blend of pure sensuousness and pure thought, exteriority and interiority, art for Hegel is "the spirit appearing in the sensuous."[9]

Of sight and hearing, however, hearing is the most ideal sense. Indeed, Hegel maintains that hearing both transcends and incorporates within itself all the other senses, including sight. The ear brings aesthetic apprehension to a higher dialectical level. It is the ear, and the ear only, that may establish the ideal correspondence between inner subjectivity of the perceiver and the spiritual interiority of the object perceived. In this way, the perceiving subject receives and so in a sense corresponds to the object whose ideal, because spiritual, interior is mediated by the sounds it emits.[10] Unlike the eye, then, the ear succeeds in apprehending both material objectivity and interiority, all at once.

Such an idealist theory of aesthetic perception is circumscribed by a long philosophical tradition—the metaphysics of presence. Consequently, it is also marked by a certain historicity. Discussing Hegel's hierarchy of the senses, Jacques Derrida suggests that Hegel could not imagine the machine, that is, a machine that functions by itself and that does not work in the service of meaning—*au service du sens*—but rather in the service of exclusive exteriority and endless repetition.[11] Derrida does not state it explicitly, but it is clear that after the advent of devices for reproducing sound, the sense of hearing can no longer be thought of as a priori ideal. A device such as the phonograph strips sound of what Hegel would call its soulful interiority, and the sensory experience of acoustic phenomena henceforth has to resort to a stubborn and ever-reproducible exteriority. The same is true of sight: its assumed ideality is exploded in the wake of inventions such as photographic means of recording visual data. In short, from now on the potentially sublime operations of the eye and the ear know an internal cleavage.

Significantly, post-Hegelian theories of aesthetic perception increasingly tend to locate the sublime, and hence also the aesthetic itself, not in the ideal realm but in the corporeal. And this bodily realm is no longer necessarily of a generalized, transcendental order, as in the aesthetic theories of Baumgarten, Kant, and Hegel. In short, the aesthetic body is no longer a universal notion. Rather, the aesthetic now tends to be located in a particular body, concrete, singular, and finite. Proust offers a pertinent example. In *Remembrance,* it is the individual body, concrete and emphatically singular, that mediates between past and present. The five senses ensure that history intersects with identity. It falls to the senses, and to them alone, to secure genuine experience. The infrastructure of the physical organism is at the core of Proust's aesthetic program.

Materialist theories of the aesthetic can be traced in numerous modernist texts from the early twentieth century onward, but perhaps nowhere so explicitly and so strongly as in Roland Barthes's book on photography, *Camera Lucida* (1980). Articulating something like an aesthetics of visual perception, Barthes dwells on the experience of looking at photographs. In an effort to pinpoint what is characteristically photographic about photography, he introduces the notion of *punctum*. A chance experience, *punctum* cannot be enforced. Rigorously singular, it is prior to language and cognition. Thus construed as an event that resists generality, habitual perception, and polite interest, *punctum* signifies a sublime form of aesthetic experience deriving from the physiological immanence of the spectator. So it is that Barthes inverts the hierarchy of body and mind inherent in Western philosophy, including its theories of the aesthetic: "I wanted to explore it," he writes apropos of the photographic image, "not as a question (a theme) but as a wound: I see, I feel, hence I notice, I observe, and I think." Perception, grounded in the body, is thus anterior to reflection. Ergo: I exist not so much because I think, but rather because I perceive. It is a perception that hurts and must hurt: "A photograph's *punctum* is that accident which pricks me (but also bruises me, is poignant to me)."[12] When *punctum* happens to the spectator, the body is taken hostage, an experience associated less with pleasure and more with pain. Barthes thus posits the body as the touchstone for the photographic *punctum;* indeed, the immanence of the body mediates experiential autonomy and authenticity. Not any body, however. In keeping with the ideal of singularity, Barthes invokes his own particular body.

Barthes's aesthetics of corporeal immediacy, like that of Proust and Joyce, is enabled and inscribed by a historically specific discourse where the materiality of the body has become the privileged site of aesthetics and where perception is an aesthetically gratifying activity in its own right. And such a discourse, as I have argued, becomes possible in the period that sees the emergence of technologies for exploring and reproducing the visual and audible real. In tracing this dialectic across three quintessentially modernist texts, we have glimpsed the historical originality that animates the novels of Mann, Proust, and Joyce. We have seen how the discursive environments of a variety of technoscientific configurations are a vital part of the inner logic of these works, including their narrative structures, thematic obsessions, and formal procedures. A sustained reconsideration of aesthetic modernism thus has to include a historically reflexive and multileveled attempt at incorporating the operations of the perceptual technologies of the second machine age into an understanding of the modernist enterprise. This, as I have attempted to show, yields a richer conception of modernist aesthetics, enhancing our understanding of the boldness of its formal innovations, the heterogeneity of its historical beginnings, and the grandeur of its sublime successes and inevitable failures.

Notes

Introduction

1. Friedrich Nietzsche, *Human, All Too Human: A Book for Free Spirits*, trans. R. J. Hollingdale (Cambridge: Cambridge University Press, 1986), 378.

2. On the cultural "assimilation of the machine" in the modernist period, see Lewis Mumford, *Technics and Civilization* (New York: Harcourt, Brace, 1934), 321–63. For a survey of the cultural consequences of technological change, see Reyner Banham, *Theory and Design in the First Machine Age*, 2d ed. (Cambridge, Mass.: MIT Press, 1980). For a wide-ranging cultural study of the advent of electricity, see Carolyn Marvin, *When Old Technologies Were New: Thinking about Electric Communication in the Late Nineteenth Century* (New York: Oxford University Press, 1988).

3. According to Andreas Huyssen, for example, "no other single factor has influenced the emergence of the new avantgarde art as much as technology, which not only fueled the artists' imagination [. . .], but penetrated to the core of the work itself" ("The Hidden Dialectic: Avantgarde—Technology—Mass Culture," in *After the Great Divide: Modernism, Mass Culture, Postmodernism* [Bloomington: Indiana University Press, 1986], 9). On avant-garde culture and technology, see, for example, R. L. Rutsky, "The Avant-Garde Techne and the Myth of Functional Form," in *High Technē* (Minneapolis: University of Minnesota Press, 1999), 73–101.

4. On the discrepancy between seeing and knowing in the wake of photography and cinema, see Karen Jacobs's study of modernist representations of the observer, *The Eye's Mind: Literary Modernism and Visual Culture* (Ithaca, N.Y.: Cornell University Press, 2001).

5. Richard Wagner himself spoke of the united artwork of the future; see Wagner, "The Art-Work of the Future," in *The Art-Work of the Future and Other Works*, trans. William Ashton Ellis (Lincoln: University of Nebraska Press, 1993), 69–213. In this essay, incidentally, Wagner sets up an opposition between art and the machine (85).

6. Fredric Jameson's designation; see Jameson, *Postmodernism; or, The Cultural Logic of Late Capitalism* (Durham, N.C.: Duke University Press, 1991), 305. "Modern art," Jameson writes, "drew its power and possibilities from being a backwater and an archaic holdover within a modernizing economy: it glorified, celebrated, and drama-

tized older forms of individual production which the new mode of production was elsewhere on the point of displacing and blotting out. [. . .] As a form of production, then, modernism (including the Great Artists and producers) gives off a message that has little to do with the content of the individual works: it is the aesthetic as sheer autonomy, as the satisfactions of handicraft transfigured" (307).

7. On the relations of modernism and postmodernism, specifically how the history of the modernist project may be rethought in the wake of the postmodern, see Jameson, *Postmodernism*, 297–418. For a reconsideration of visual modernism, see T. J. Clark, *Farewell to an Idea: Episodes from a History of Modernism* (New Haven, Conn.: Yale University Press, 1999). For a materialist perspective on the institutions that made possible so many high-modernist writing careers, predominantly in a British context, see Lawrence Rainey, *Institutions of Modernism: Literary Elites and Public Cultures* (New Haven, Conn.: Yale University Press, 1998), esp. the chapter on Joyce, 42–76. For a feminist perspective on literary modernism, see Sandra M. Gilbert and Susan Gubar, *No Man's Land: The Place of the Woman Writer in the Twentieth Century*, 3 vols. (New Haven, Conn.: Yale University Press, 1988–94).

8. An early example is Stephen Kern's pioneering work on modernism and technology, *The Culture of Time and Space, 1880–1918* (Cambridge, Mass.: Harvard University Press, 1983). For a wide-ranging cultural study of how modernity relates to modernism, see Marshall Berman, *All That Is Solid Melts into Air: The Experience of Modernity* [1982] (New York: Penguin, 1988). On how the notion of literature is affected by the advent of technologies for representing analogically the real, primarily in a German context, see Friedrich A. Kittler, *Discourse Networks, 1800/1900*, trans. Michael Metteer, with Chris Cullens, with a foreword by David E. Wellbery (Stanford, Calif.: Stanford University Press, 1990); a revised and augmented version of the German original appeared in 1995. I say more about these works in Chapter 1. On how modernist texts reflect changing conceptions of the body in the wake of technoscientific change, particularly in an Anglo-American context, see Tim Armstrong, *Modernism, Technology, and the Body: A Cultural Study* (Cambridge: Cambridge University Press, 1998). Two excellent literary studies address American modernism; see Cecelia Tichi, *Shifting Gears: Technology, Literature, Culture in Modernist America* (Chapel Hill: University of North Carolina Press, 1987); and Lisa M. Steinman, *Made in America: Science, Technology, and American Modernist Poets* (New Haven, Conn.: Yale University Press, 1987). See also Mark Seltzer's study of the body/machine complex in American realism and naturalism, *Bodies and Machines* (New York: Routledge, 1992). For a study of scientific management and early-twentieth-century American narrative culture, see Martha Banta, *Taylored Lives: Narrative Productions in the Age of Taylor, Veblen, and Ford* (Chicago: University of Chicago Press, 1993). On the cultural significance of Edison's inventions, specifically in an American context, see Lisa Gitelman, *Scripts, Grooves, and Writing Machines* (Stanford, Calif.: Stanford University Press, 1999). See also Rutsky, *High Technē;* although Rutsky does not address literary modernism as such, the book is a useful overview of how technological change affects aesthetic innovation in the twentieth century.

9. Italo Calvino, *Six Memos for the Next Millennium*, trans. Patrick Creagh (Cambridge, Mass.: Harvard University Press, 1988), 112.

10. Jameson, *Postmodernism*, 304.

11. I am here influenced by Wolfgang Schivelbusch and Lynne Kirby, who approach

the railroad car as a mobile viewing instrument, and by Paul Virilio, who approaches aviation as a way of seeing. Their perspectives may usefully be extended to the automobile, as my discussion of Proust will make clear. See Schivelbusch, *The Railway Journey: The Industrialization of Time and Space in the Nineteenth Century* (Berkeley: University of California Press, 1986); Kirby, *Parallel Tracks: The Railroad and Silent Cinema* (Durham, N.C.: Duke University Press, 1997); and Virilio, *War and Cinema: The Logistics of Perception*, trans. Patrick Camiller (London: Verso, 1989).

12. On modern discourses of vision and visuality, see Martin Jay, *Downcast Eyes: The Denigration of Vision in Twentieth-Century French Thought* (Berkeley: University of California Press, 1993); and David Michael Levin, ed., *Modernity and the Hegemony of Vision* (Berkeley: University of California Press, 1993).

13. Terry Eagleton, *The Ideology of the Aesthetic* (Oxford: Basil Blackwell, 1990), 13, 13–17.

14. For a full etymological explanation, see H. G. Liddell and R. Scott, *Greek-English Lexicon*, 9th ed. (Oxford: Oxford University Press, 1996).

15. I am paraphrasing Barbara Herrnstein Smith's formula in *Contingencies of Value: Alternative Perspectives for Critical Theory* (Cambridge, Mass.: Harvard University Press, 1988), 40.

16. On these differences, see Matei Calinescu, *Five Faces of Modernity: Modernism, Avant-Garde, Decadence, Kitsch, Postmodernism*, 2d rev. ed. (Durham, N.C.: Duke University Press, 1987), 68–85. On the differences between a French conception of modernism and an Anglo-American one, see Antoine Compagnon, *The Five Paradoxes of Modernity*, trans. Franklin Philip (New York: Columbia University Press, 1994).

17. See Harry Levin, "What Was Modernism?" in *Refractions: Essays in Comparative Literature* (New York: Oxford University Press, 1966), 283.

18. For a comprehensive chronology of events, see *Modernism, 1890–1930*, ed. Malcolm Bradbury and James McFarlane (London: Penguin, 1976), 571–612. See also the 1913–1925 chronology in *Making Mischief: Dada Invades New York*, ed. Francis M. Naumann, with Beth Venn (New York: Whitney Museum of American Art, 1996), 34–175.

19. On the dialectics of mass culture and high modernism, see Raymond Williams, "Culture and Technology," in *The Politics of Modernism: Against the New Conformists*, ed. Tony Pinkney (London: Verso, 1989), 119–39; Fredric Jameson, "Reification and Utopia in Mass Culture," in *Signatures of the Visible* (New York: Routledge, 1990), 9–54; Franco Moretti, "Modern European Literature: A Geographical Sketch," *New Left Review* 206 (July–August 1994): 86–109. See also John Carey, *The Intellectual and the Masses: Pride and Prejudice among the Literary Intelligentsia, 1880–1939* (London: Faber and Faber, 1992).

20. Huyssen, *After the Great Divide*, viii.

21. See also Armstrong, *Modernism, Technology, and the Body*, 4.

22. Junichiro Tanizaki, *In Praise of Shadows*, trans. Thomas J. Harper and Edward G. Seidensticker (London: Jonathan Cape, 1991), 24.

23. For a discussion of this concept, see Michael Hardt, *Gilles Deleuze: An Apprenticeship in Philosophy* (Minneapolis: University of Minnesota Press, 1993), 2–10, 97–100.

24. In using the word "technoscientific," which derives from Bruno Latour, I want to signal that no absolute distinction between technology and science can be made. See Bruno Latour, *Science in Action: How to Follow Scientists and Engineers through Society* (Cambridge, Mass.: Harvard University Press, 1987), 174–75. I should stress, too,

that a rich body of work explores the relations of science and literature. See, for example, Michel Serres, *Hermes: Literature, Science, Philosophy,* ed. Josué V. Harari and
David F. Bell (Baltimore, Md.: Johns Hopkins University Press, 1982); and N. Katherine Hayles, *The Cosmic Web: Scientific Field Models and Literary Strategies in the Twentieth Century* (Ithaca, N.Y.: Cornell University Press, 1984). On the affinities between
science and modernist art, see Linda Dalrymple Henderson, *The Fourth Dimension
and Non-Euclidean Geometry in Modern Art* (Princeton, N.J.: Princeton University
Press, 1983), as well as her *Duchamp in Context: Science and Technology in "The Large
Glass" and Related Works* (Princeton, N.J.: Princeton University Press, 1998).

25. In the meantime, see, for example, Huyssen, "Mass Culture as Woman: Modernism's Other," in *After the Great Divide,* 44–62; Kittler, "Queen's Sacrifice," in *Discourse Networks,* 347–68; Armstrong, "Making a Woman," in *Modernism, Technology,
and the Body,* 159–83; Katherine Stubbs, "Mechanizing the Female: Discourse and Control in the Industrial Economy," *differences* 7, no. 1 (1995): 141–64; Michael Wutz, "The
Thermodynamics of Gender: Lawrence, Science and Sexism," *Mosaic* 28, no. 2 (1995):
83–108; and Rutsky's reading of Fritz Lang's *Metropolis,* "The Mediation of Technology
and Gender," in *High Technē,* 48–72.

26. Karl Marx, "Economic and Philosophical Manuscripts," in *Early Writings,*
trans. Rodney Livingstone and Gregor Benton (London: Penguin, 1984), 353.

27. See Guy Debord, *The Society of the Spectacle* (Detroit: Black and Red, 1983); Paul
Virilio, *War and Cinema,* as well as his *The Vision Machine,* trans. Julie Rose (Bloomington: Indiana University Press, 1994); Marshall McLuhan, *Understanding Media: The
Extensions of Man,* with an introduction by Lewis H. Lapham (Cambridge, Mass.:
MIT Press, 1994); Kittler, *Discourse Networks,* part 2, 177–368, as well as his *Gramophone, Film, Typewriter,* trans., with an introduction, by Geoffrey Winthrop-Young
and Michael Wutz (Stanford, Calif.: Stanford University Press, 1999).

28. Jean-Yves Tadié has stressed the role of modern technology in *Remembrance;*
see Tadié, *Proust: Biographie* (Paris: Gallimard, 1996), 599–600.

29. That Proust may be seen as a theorist has been suggested by Malcolm Bowie,
who maintains that Proust's tale is "one of the most elaborate and circumstantial por-
trayals of the theorising mind that European culture possesses" (*Freud, Proust and
Lacan: Theory as Fiction* [Cambridge: Cambridge University Press, 1987], 65). Moreover, Siegfried Kracauer, in *Theory of Film,* approaches Proust as theorist of photography, basing his ontology of the photographic image in an analysis of Proust's camera-
eye episode; see Kracauer, *Theory of Film: The Redemption of Physical Reality*
(Princeton, N.J.: Princeton University Press, 1960), esp. 14–20. In Kracauer's last work,
Proust's camera-eye episode also plays an important role; see *History: The Last Things
before the Last* (New York: Oxford University Press, 1969), 49–52, 82–86, 92–93. For a
discussion of Kracauer's reading of Proust, see Angela Cozea, "Proustian Aesthetics:
Photography, Engraving, and Historiography," *Comparative Literature* 45, no. 1 (winter
1993): 209–29. Alan Spiegel, finally, uses Proust's camera-eye episode as a theoretical
point of departure in his discussion of preeminently cinematic features in modernist
fiction; see Spiegel, *Fiction and the Camera Eye: Visual Consciousness in Film and the
Modern Novel* (Charlottesville: University Press of Virginia, 1976), esp. 82–89.

30. Proust's telephone episode has a rich prehistory. An early version appears in
Jean Santeuil, relating Jean's first telephone conversation with his mother. In 1907, just

about a decade later, Proust published in *Le Figaro* a modified version in an essay on reading. Having revised the episode once again, he made it a part of *The Guermantes Way*. See Paul Martin, "Le téléphone: Etude littéraire d'un texte de M. Proust," parts 1–3, *Information littéraire* 21 (1969): 233–41; 22 (1970): 46–52, 87–98. For an inventory of the cultural imaginary of the telephone in Proust's time, see *Le téléphone à la Belle Epoque* (Brussels: Editions Libro-Sciences, 1976).

31. The Orpheus myth has typically been read as a story about the deadly gaze, but as Proust's telephone episode intimates, it is also a tale about the faculty of listening and the animating power of the voice. On this parallel, see my "Orpheus and the Machine," *Forum for Modern Language Studies* 37, no. 2 (2001): 127–40.

32. I stress this point because Proust's episode shares an affinity with Benjamin's notion of the aura. Benjamin approaches aura in two ways: in terms of spatiotemporal uniqueness, and in terms of the gaze. In mechanically reproducing the visual real, the photographic image strips the object of its unique presence in time and space; at the same time, photography makes the past look at us, but—and this is Benjamin's vital point—we cannot look back. For this reason, photography is linked to death. It is not the gaze itself that is deadly, however; it is the failure to meet the gaze of the other that is deadly. Benjamin's history of the decline of aura is also the history of an increasing inability to meet the intentional and unique gaze of the other, be it an object, a human being, or history itself. See Benjamin, "A Small History of Photography," in *One-Way Street and Other Writings*, trans. Edmund Jephcott and Kingsley Shorter (London: NLB, 1979), 240–57; "The Work of Art in the Age of Mechanical Reproduction," in *Illuminations*, ed. Hannah Arendt, trans. Harry Zohn (New York: Schocken, 1988), 217–51; and "Some Motifs in Baudelaire," in *Charles Baudelaire: A Lyric Poet in the Era of High Capitalism*, trans. Harry Zohn (London: Verso, 1973), 109–54.

33. McLuhan, *Understanding Media*, 64.

34. For a cultural history of the gramophone and its impact on acoustic representation, see Kittler, *Gramophone, Film, Typewriter*, 21–114, as well as his *Discourse Networks*, 229–64. For an interesting attempt to theorize the historical construction of acoustic experience in the late nineteenth century, see Jay Clayton, "The Voice in the Machine: Hazlitt, Hardy, James," in *Language Machines: Technologies of Literary and Cultural Production*, ed. Jeffrey Masten, Peter Stallybrass, and Nancy J. Vickers (New York: Routledge, 1997), 209–32.

35. Virginia Woolf, *Mrs Dalloway* (New York: Modern Library, 1928), 19.

36. Robert Musil, *The Man without Qualities*, trans. Sophie Wilkins and Burton Pike, 2 vols. (New York: Knopf, 1995), 2:785.

37. Wyndham Lewis, *Tarr. The 1918 Version*, ed. Paul O'Keeffe (Santa Rosa, Calif.: Black Sparrow Press, 1990), 99–100.

38. See Jonathan Crary, *Techniques of the Observer: On Vision and Modernity in the Nineteenth Century* (Cambridge, Mass.: MIT Press, 1990). On how modernist fiction, from Woolf and Ellison to Nabokov, reflects the emergence of the embodied observer, see Jacobs, *The Eye's Mind*.

39. Armstrong, *Modernism, Technology, and the Body*, 2.

40. Joseph Conrad, preface to *The Nigger of the "Narcissus"* (London: Penguin, 1988), xlix.

41. Conrad, *Lord Jim* (London: Penguin, 1986), 56. For an analysis of Conrad's writerly style as an essentially "aestheticizing strategy," see Jameson, *The Political Un-*

conscious: Narrative as Socially Symbolic Act (Ithaca, N.Y.: Cornell University Press, 1981), 229–33. On modernist literary techniques of visual representation, see Diane R. Leonard, "Proust and Virginia Woolf, Ruskin and Roger Fry: Modernist Visual Dynamics," in *Comparative Literature Studies* 18, no. 3 (September 1981): 333–43.

42. Ezra Pound, "Machine Art," in *Machine Art and Other Writings: The Lost Thought of the Italian Years*, ed. Maria Luisa Ardizzone (Durham, N.C.: Duke University Press, 1996), 77. On Pound's notion of *techne* and *poiesis*, see Ardizzone's introduction to *Machine Art*, 18–25.

1. The Antitechnological Bias and Other Modernist Myths

1. Louis Aragon, *Nightwalker*, trans. Frederick Brown (Englewood Cliffs, N.J.: Prentice-Hall, 1970), 163.

2. Malcolm Bradbury and James McFarlane, "The Name and Nature of Modernism," in *Modernism, 1890–1930* (1976; London: Penguin, 1983), 27.

3. Bruno Latour makes this point in *We Have Never Been Modern*, trans. Catherine Porter (Cambridge, Mass.: Harvard University Press, 1993), 8.

4. Strong readings, Bloom suggests, are always misreadings: "strong poets [. . .] make [poetic history] by misreading one another, so as to clear imaginative space for themselves" (*The Anxiety of Influence: A Theory of Poetry* [London: Oxford University Press, 1973], 5).

5. Slavoj Žižek, "Alfred Hitchcock, or, The Form and Its Historical Mediation," in *Everything You Always Wanted to Know about Lacan (But Were Afraid to Ask Hitchcock)*, ed. Slavoj Žižek (London: Verso, 1992), 1, 2, 2.

6. On the "exile of evaluation," particularly in Anglo-American scholarship and criticism, see Barbara Herrnstein Smith, *Contingencies of Value: Alternative Perspectives for Critical Theory* (Cambridge, Mass.: Harvard University Press, 1988), 17–29.

7. See notes 7 and 8 in the introduction to the present work.

8. Edward Said, *Culture and Imperialism* (New York: Knopf, 1993), 188. Said's suggestion finds implicit support in Moretti's speculative essay "Modern European Literature: A Geographical Sketch," *New Left Review*, no. 206 (July-August 1994): 86–109. In developing a spatial, geographical model for understanding the history of European literature since the seventeenth century, Moretti paves the way for a proto-geopolitical analysis of the significance of this history: "In the spatial model [. . .] geography is no longer the speechless onlooker of the—historical—deeds of the 'European spirit'. The European space is not a landscape, not backdrop of history, but a *component* of it" (92). On the conjuncture between modernism and imperialism, see Fredric Jameson, "Modernism and Imperialism," in *Nationalism, Colonialism, and Literature*, ed. Seamus Deane (Minneapolis: University of Minnesota Press, 1990), 43–66; and Edward Said, "Yeats and Decolonization," in *Nationalism, Colonialism, and Literature*, 69–95. For related perspectives on primitivism in the modernist period, see Marianna Torgovnick, *Gone Primitive: Savage Intellects, Modern Lives* (Chicago: University of Chicago Press, 1990).

9. Paul Valéry, "La crise de l'esprit," in *Variété*, vol. 4, bk. 1 of *Œuvres* (Paris: Editions de la NRF, 1934), 13.

10. Ernst Robert Curtius argues that the use of the word *modernus* appeared only

in the sixth century, whereas Matei Calinescu, basing his argument on more recent research, holds that the word was already in use in the late fifth century; see Curtius, *European Literature and the Latin Middle Ages*, trans. Willard Trask (London: Routledge and Kegan Paul, 1953), 19–20, 251–53; and Calinescu, *Five Faces of Modernity: Modernism, Avantgarde, Decadence, Kitsch, Postmodernism.* 2d ed., rev. (Durham, N.C.: Duke University Press, 1993), 13–14.

11. See, for example, Calinescu, *Five Faces of Modernity;* and Frederick R. Karl, *Modern and Modernism: The Sovereignty of the Artist, 1885–1925* (New York: Atheneum, 1985), 3–39. See also Williams, "When Was Modernism?" in *The Politics of Modernism,* ed. Tony Pinkney (London: Verso, 1989), 32.

12. I am here excluding accounts of the avant-garde.

13. This is the place to underscore that the critique that follows is anything but a wholesale dismissal of the works under discussion. I owe numerous significant insights into modernist art and culture to these works of criticism.

14. Douwe Fokkema, *Literary History, Modernism, and Postmodernism* (Amsterdam: John Benjamins, 1984), 34.

15. But see also Douwe Fokkema and Elrud Ibsch, *Modernist Conjunctures: A Mainstream in European Literature, 1910–1940* (London: Hurst, 1987), esp. 1–47, where the authors attempt to combine the essentialist approach and the idea of a specifically modernist semantic universe with reception theory.

16. Hugo Friedrich, *The Structure of Modern Poetry: From the Mid-Nineteenth to the Mid-Twentieth Century,* trans. Joachim Neugroschel (Evanston, Ill.: Northwestern University Press, 1974), 129, 161.

17. For a critical discussion of Greenberg's formalist notion of the modernist work of art, see, for example, Johanna Drucker, *Theorizing Modernism: Visual Art and the Critical Tradition* (New York: Columbia University Press, 1994), 82–90.

18. Clement Greenberg, "Avant-Garde and Kitsch," in *The Collected Essays and Criticism,* ed. John O'Brian (Chicago: University of Chicago Press, 1986), 1:15.

19. Calinescu, *Five Faces of Modernity,* 41. *Five Faces* is an enlarged, revised version of *Faces of Modernity: Avant-Garde, Decadence, Kitsch* (1977), the essential difference being that the second edition includes a chapter on postmodernism. Contrary to what one might have expected, however, literary modernism is not subjected to critical treatment as a "face of modernity."

20. For other accounts of modernism in which the opposition between the elite and the masses, art and mass culture, is an organizing feature, see José Ortega y Gasset, "The Dehumanization of Art," in *The Dehumanization of Art,* trans. Helene Weyl (Princeton, N.J.: Princeton University Press, 1968), 3–50; and Harry Levin, "What Was Modernism?" in *Refractions: Essays in Comparative Literature* (New York: Oxford University Press, 1966), 271–95.

21. Calinescu, *Five Faces,* 42.

22. Ricardo J. Quinones, *Mapping Literary Modernism: Time and Development* (Princeton, N.J.: Princeton University Press, 1985), 66, 65, 67, 66.

23. Astradur Eysteinsson, *The Concept of Modernism* (Ithaca: Cornell University Press, 1990), 16.

24. See Raymond Williams, "When Was Modernism?" and "Culture and Technology," in *Politics of Modernism,* 31–35, 119–39; Kern, *Culture of Time and Space;* Berman,

All That Is Solid Melts into Air: The Experience of Modernity (New York: Penguin, 1988); and Benjamin, "The Work of Art in the Age of Mechanical Reproduction" and "The Storyteller," in *Illuminations*, trans. Harry Zohn (New York: Schocken, 1988), 217–51, 83–109, as well as his "Author as Producer," in *Reflections*, ed. Peter Demetz, transl. Edmund Jephcott (New York: Schocken, 1988), 220–238.

25. See, for example, Jacques Derrida, "Economimesis," trans. R. Klein, *Diacritics* 11, no. 2 (June 1981): 3–25. Similary, as Jeffrey Masten et al. point out, early uses of words such as "machine" and "mechanical" are closely connected to the semantic sphere of the body and the flesh; see the introduction to *Language Machines: Technologies of Literary and Cultural Production*, ed. Jeffrey Masten, Peter Stallybrass, and Nancy J. Vickers (New York: Routledge, 1997), 1–2.

26. Theodor W. Adorno and Max Horkheimer, *Dialectic of Enlightenment*, trans. John Cumming (London: Verso, 1989), 126.

27. See Williams, *Culture and Society*, esp. "The Romantic Artist," 48–64.

28. Edward Young, *Conjectures on Original Composition* (1759); quoted in Williams, *Culture and Society*, 54.

29. John Milton, *Paradise Lost*, in *The Poetical Works of John Milton* (London: T. J. Allman, 1822), 1:18.

30. For a cultural history of the technologies of the first machine age, see Humphrey Jennings's rich source book, *Pandaemonium, 1600–1886: The Coming of the Machine as Seen by Contemporary Observers*, ed. Mary-Lou Jennings and Charles Madge (New York: Free Press, 1985).

31. Martha Woodmansee, *The Author, Art, and the Market: Rereading the History of Aesthetics* (New York: Columbia University Press, 1994), esp. "The Interests in Disinterestedness," 11–33, and "Genius and the Copyright," 35–55.

32. Adorno and Horkheimer, *Dialectic of Enlightenment*, 158. For a critique of nonutilitarian aesthetics, see Barbara Herrnstein Smith, *Contingencies of Value*, 125–49.

33. Roger Chartier, *The Order of Books: Readers, Authors, and Libraries between the Fourteenth and Eighteenth Centuries*, trans. Lydia G. Cochrane (Stanford, Calif.: Stanford University Press, 1994), 37. Pierre Bourdieu similarly maintains that the ideal of the "pure" perception of a work of art *qua* work of art should be understood as "the product of the enunciation and systematization of the principles of specifically aesthetic legitimacy which accompany the constituting of a relatively autonomous artistic field" (*Distinction: A Social Critique of the Judgement of Taste*, trans. Richard Nice [Cambridge, Mass.: Harvard University Press, 1984], 30).

34. Renato Poggioli, *The Theory of the Avant-Garde*, trans. Gerald Fitzgerald (Cambridge, Mass.: Harvard University Press, 1986), 138.

35. Adorno, letter to Walter Benjamin, 18 March 1936, in *Aesthetics and Politics*, ed. Perry Anderson (London: Verso, 1986), 123.

36. In Aristotle's *Nicomachean Ethics*, efficient causality is but one among the four causal modalities discussed, and Heidegger proceeds to reinstate the fourth, *causa finalis*. This fourth category usefully challenges the received idea of the linear chronology supposed to inform causal relations. Thus Heidegger reminds his reader that the end that determines the means may also be considered a cause. From a conventional point of view, then, the cause is postponed, although its effects already make them-

selves felt. See Heidegger, "The Question Concerning Technology," in *Basic Writings*, trans. and ed. David Farrell Krell (New York: Harper Collins, 1993), 313.

37. R. L. Rutsky has similarly observed that technology tends to be seen in terms of an "instrumental rationality"; see Rutsky, *High Technē: Art and Technology from the Machine Aesthetic to the Posthuman* (Minneapolis: University of Minnesota Press, 1999), 73.

38. Williams, "Culture and Technology," 132.

39. For Bruno Latour's critique of technological determinism, see *Science in Action: How to Follow Scientists and Engineers through Society* (Cambridge, Mass.: Harvard University Press, 1987). Hereafter page references are given in the main text. See also Latour's case study of Louis Pasteur, *The Pasteurization of France*, trans. Alan Sheridan and John Law (Cambridge, Mass.: Harvard University Press, 1993), 3–150. For an anthropological study of the role of inscription machines in the scientific laboratory, see Latour and Woolgar, *Laboratory Life: The Social Construction of Scientific Facts* (Beverly Hills, Calif.: Sage Publications, 1979). For a variety of perspectives on scientific practice, see the articles collected in *Science as Practice and Culture*, ed. Andrew Pickering (Chicago: University of Chicago Press, 1992). For Donna Haraway's theory of technology, see *Simians, Cyborgs, and Women: The Reinvention of Nature* (New York: Routledge, 1991).

40. Gilles Deleuze, *Foucault*, trans. Séan Hand (Minneapolis: University of Minnesota Press, 1988), 39.

41. The telephone network does more than visualize the whole. The metaphor can be said to hint at the historicity of the problem as such. Just as telecommunications is a recent phenomenon from a historical point of view, so the difficulty of thinking the totality is in some sense a historically specific problem.

42. For analytic purposes, however, it is essential to retain distinctions between the social, the technological, the scientific, the economic, and the political, no matter how arbitrary or preliminary these distinctions might be, not because of the alleged *essence* of this or that activity, this or that realm (be it policy-making, test tubes, scientific writing, or grant bodies), but rather because of their specific and different *effects* in the realm of the human.

43. Berman, *All That Is Solid Melts into Air*.

44. For a critical discussion of Berman's study, see Perry Anderson, "Modernity and Revolution," *New Left Review*, no. 144 (March–April 1984): 96–113. Challenging Berman's ahistorical notion of modernism as worldview, Anderson proposes that modernism is an "intersection of different historical temporalities," thus to be understood as a "cultural field of force *triangulated* by three decisive coordinates": academicism, new technologies and, in his terms, the imaginative proximity of social revolution. See also Berman's rejoinder, "The Signs in the Street: A Response to Perry Anderson," *New Left Review*, no. 144 (March–April 1984): 114–23.

45. Friedrich A. Kittler, *Discourse Networks, 1800/1900*, trans. Michael Metteer, with Chris Cullens, with a foreword by David E. Wellbery (Stanford, Calif.: Stanford University Press, 1990). See also Kittler's companion volume to *Discourse Systems*, titled *Gramophone, Film, Typewriter*, trans. Geoffrey Winthrop-Young and Michael Wutz (Stanford, Calif.: Stanford University Press, 1999).

46. Winthrop-Young and Wutz touch on these critiques in their introduction to the English translation of Kittler's *Gramophone, Film, Typewriter*, esp. xxxiv–xxxv.

47. My usage of the terms "allegory" and "symbol" reflects contemporary discussions, and derives primarily from Benjamin's and de Man's respective reconsiderations of allegorical modes of signification; see Benjamin, *The Origin of German Tragic Drama*, trans. John Osborne (London: Verso, 1990); and de Man, "The Rhetoric of Temporality," in *Blindness and Insight: Essays in the Rhetoric of Contemporary Criticism*, 2d ed., rev. (London: Methuen, 1986), 187–228; as well as de Man's "Allegory (*Julie*)," in *Allegories of Reading: Figural Language in Rousseau, Nietzsche, Rilke, and Proust* (New Haven, Conn.: Yale University Press, 1979), 188–220. For a historicizing perspective on the revival of allegory, see Jameson, *Postmodernism; or, the Cultural Logic of Late Capitalism* (Durham, N.C.: Duke University Press, 1990), 167–75.

48. Hugh Kenner, *The Mechanic Muse* (Oxford: Oxford University Press, 1987), 36.

49. George P. Landow, *Hypertext: The Convergence of Contemporary Critical Theory and Technology* (Baltimore: Johns Hopkins University Press, 1992).

50. Benjamin, "Work of Art," 233.

51. Kenner, *Mechanic Muse*, 10–11.

52. Wolfgang Schivelbusch, *The Railway Journey: The Industrialization of Time and Space in the Nineteenth Century* (Berkeley: University of California Press, 1986), 159.

53. N. Katherine Hayles, "Boundary Disputes: Homeostasis, Reflexivity, and the Foundations of Cybernetics," *Configurations* 2, no. 3 (1994): 464.

54. Aristotle, *The Poetics*, in S. H. Butcher, ed., *Aristotle's Theory of Poetry and Fine Art*, 4th ed. (New York: Dover Publications, 1951), 80–81.

55. Susan Buck-Morss, *The Dialectics of Seeing: Walter Benjamin and the Arcades Project* (Cambridge, Mass.: MIT Press, 1989), 124–45, quotation, 124.

56. As is well known, Benjamin's Baudelaire study grew out of two exposés for *The Arcades Project*, "Paris, the Capital of the Nineteenth Century" (1935) and "Paris, Capital of the Nineteenth Century" (1939). See Benjamin, *The Arcades Project*, trans. Howard Eiland and Kevin McLaughlin, prepared on the basis of the German volume edited by Rolf Tiedemann (Cambridge, Mass.: Belknap Press/Harvard University Press, 1999), 3–13, 14–26. See also Benjamin's convolute on Baudelaire in *The Arcades Project*, 228–387. On the history of Benjamin's Baudelaire essay and its significance for *The Arcades Project*, see Buck-Morss, *Dialectics of Seeing*, 205–10. See also Buck-Morss, *The Origin of Negative Dialectics: Theodor W. Adorno, Walter Benjamin, and the Frankfurt Institute* (New York: Free Press, 1977), 155–63.

57. Like numerous accounts of the interrelations of culture and technology, however, Benjamin's reading projects a rudimentary historical narrative that posits an originary, or organic, moment and a subsequent, disintegrated moment. Whether the object of study is the emergence of scriptural culture, as in Walter Ong's *Orality and Literacy: The Technologizing of the Word* (1982), the printing revolution in the West in the fifteenth century, as in Elisabeth Eisenstein's *The Printing Press as an Agent of Change* (1979), the impact of the railway upon human consciousness, as in Wolfgang Schivelbusch's *Railway Journey*, or machine-powered technology as *Gestell*, as in Martin Heidegger's "The Question of Technology" (1953), the implied historiographic structure tends to assume the form of the myth of the Fall. This conceptual design is particularly striking in studies of modernism and modernity, where "organic," or "traditional," or "premodern" society is usually posited against the modern. Although such a structure may be understood as a necessary heuristic device, it is essential to de-

velop a more differentiated theoretical and historical framework, where the social, economic, and cultural transformation that is so often designated as a transition from *Gemeinschaft* to *Gesellschaft* is both accounted for and problematized as a lingering conceptual structure linked to the self-understanding of the moderns.

58. Benjamin, "N [On the Theory of Knowledge, Theory of Progress]," in *The Arcades Project*, 463.

59. That Benjamin's mode of allegorization may be seen as a vehicle for representing mediation is an idea Fredric Jameson has generously shared with me. On Benjamin's notion of allegory, see Buck-Morss, *Dialectics of Seeing*, 160–201.

60. Benjamin, "Some Motifs in Baudelaire," in *Charles Baudelaire: A Lyric Poet in the Era of High Capitalism*, trans. Harry Zohn (London: Verso, 1989), 109–54. Hereafter page references are given in the main text.

61. Jürgen Habermas, "Consciousness-Raising or Rescuing Critique," in *Walter Benjamin: Critical Essays and Recollections*, ed. Gary Smith, trans. Frederick Lawrence (Cambridge, Mass.: MIT Press, 1991), 109.

62. Adorno, "The Essay as Form," in *Notes to Literature*, trans. Shierry Weber Nicholson, 2 vols. (New York: Columbia University Press, 1991), 1:19.

2. Novel Visions and the Crisis of Culture

1. In preparing this chapter, I have consulted the following bibliographies and works of reference: Klaus W. Jonas, *Die Thomas-Mann-Literatur: Bibliographie der Kritik*, 2 vols. (Berlin: Erich Schmidt, 1972–79); Harry Matter, *Die Literatur über Thomas Mann: Eine Bibliographie, 1898–1969*, 2 vols. (Berlin: Aufbau, 1972); Herbert Lehnert, *Thomas-Mann-Forschung: Ein Bericht* (Stuttgart: Metzler, 1969); Hermann Kurzke, *Thomas-Mann-Forschung, 1969–1976: Ein kritischer Bericht* (Frankfurt am Main: S. Fischer, 1977); *Stationen der Thomas-Mann-Forschung: Aufsätze seit 1970*, ed. Hermann Kurzke (Würzburg: Königshausen und Neumann, 1985); *Bibliographie der deutschen Sprach-und Literaturwissenschaft* (Frankfurt am Main: V. Klostermann, 1970–); Hans Rudolf Vaget, *Thomas Mann: Studien zu Fragen der Rezeption* (Bern: Herbert Lang, 1975); Helmut Koopmann, ed. *Thomas-Mann-Handbuch* (Stuttgart: A. Kröner, 1990); Hugh Ridley, *The Problematic Bourgeois: Twentieth-Century Criticism on Thomas Mann's "Buddenbrooks" and "The Magic Mountain"* (Columbia, S.C.: Camden House, 1994).

2. On the relations of modernist visual arts, particularly surrealism, and technology in a more general sense, see Rosalind E. Krauss, *The Originality of the Avant-Garde and Other Modernist Myths* (Cambridge, Mass.: MIT Press, 1985), as well as her *Optical Unconscious* (Cambridge, Mass.: MIT Press, 1993). See also Hal Foster, *Compulsive Beauty* (Cambridge, Mass.: MIT Press, 1993); and Linda Dalrymple Henderson, "X Rays and the Quest for Invisible Reality in the Art of Kupka, Duchamp, and the Cubists," *Art Journal* 47, no. 4 (1988): 323–40.

3. *Aisthesis*: Greek, from *aisthetikos*, from *aisthanomai*, to perceive, apprehend by the senses; [of mental perception] perceive, understand; hear, learn; take notice of, have perception of. For a full etymological explanation, see H. G. Liddell and R. Scott, *Greek-English Lexicon* (Oxford: Clarendon, 1996).

4. Jonathan Crary, *Techniques of the Observer: On Vision and Modernity in the Nineteenth Century* (Cambridge, Mass.: MIT Press, 1990). On twentieth-century critiques of vision as a philosophical paradigm, particularly in the French context, see Martin Jay, *Downcast Eyes: The Denigration of Vision in Twentieth-Century French Thought* (Berkeley: University of California Press, 1993).

5. *Theoreo:* look at, behold; inspect, review; gaze, gape; [of the mind] contemplate, consider; observe; perceive, speculate, theorize. For a full etymological explanation, see Liddell and Scott, *Greek-English Lexicon.*

6. John Ruskin, *The Eagle's Nest: Ten Lectures on the Relation of Natural Science to Art,* in vol. 22 of *The Works of John Ruskin,* ed. E. T. Cook and Alexander Wedderburn (London: George Allen, 1906), 210. In Chapter 4, I will discuss Turner's distinction in more detail.

7. But see now Geoffrey Winthrop-Young, "Magic Media Mountain: Technology and the *Umbildungsroman,*" in *Reading Matters: Narrative in the New Media Ecology,* ed. Joseph Tabbi and Michael Wutz (Ithaca, N.Y.: Cornell University Press, 1997), 29–52. Winthrop-Young's interesting essay was published when my chapter was complete, and some of his critical perspectives may be usefully aligned with my own, particularly the focus on how "medical technology and storage technology" affect the "cultural construction of the human body," including the human sensorium (32). Influenced by Kittler's *Discourse Networks,* Winthrop-Young suggests that Mann's novel is "the first epic of modern information." Bespeaking the crisis of the novel and of culture in general, *The Magic Mountain* teaches that "books are demoted, information is circulated by more effective technologies, and those technologies now dominate, invade, and will eventually replace bodies" (50). The idea that Mann's novel thematizes the emergence of new media such as cinema has been suggested by Jochen Hörisch, " 'Die deutsche Seele up to date': Sakramente der Medientechnik auf dem Zauberberg," in *Arsenale der Seele: Literatur- und Medienanalyse seit 1870,* ed. Friedrich Kittler and Georg Christoph Tholen (Munich: Fink, 1989), 13–23.

8. It is well known that Mann underwent a political and intellectual reorientation during this period. Published in 1918, *Reflections of a Non-Political Man* is Mann's widely debated and massive testimony of his antidemocratic, romanticist, and patriotic beliefs. By the time he had finished *The Magic Mountain,* he identified himself as a liberal democrat.

9. Hans Levander, *Thomas Mann: Silhuetten och verket* (Stockholm: Bonniers, 1964), 166.

10. Italo Calvino, *Six Memos for the Next Millennium,* trans. Patrick Creagh (Cambridge, Mass.: Harvard University Press, 1988), 116.

11. Pierre Bourdieu, "The Invention of the Artist's Life," trans. Erec. R. Koch, *Yale French Studies* 73 (1987): 80.

12. For an analysis of *The Magic Mountain* as a renewal and parody of the classical bildungsroman, see Jürgen Scharfschwerdt, *Thomas Mann und der deutsche Bildungsroman* (Stuttgart: W. Kohlhammer, 1967), esp. 114–74. See also Martin Swales, *The German Bildungsroman from Wieland to Hesse* (Princeton, N.J.: Princeton University Press, 1978); Jochen Hörisch, *Gott, Geld und Glück: Zur Logik der Liebe in den Bildungsromanen Goethes, Kellers und Thomas Manns* (Frankfurt am Main: Suhrkamp, 1983); Ulrich Thomet, *Das Problem der Bildung im Werke Thomas Manns* (Bern: Her-

bert Lang, 1975), esp. 66–93. Helmut Koopmann, by contrast, argues that *The Magic Mountain* is not so much a bildungsroman as a novel of initiation or, alternatively, an intellectual novel. See Koopmann, *Die Entwicklung des "intellektualen Romans" bei Thomas Mann: Untersuchungen zur Struktur von "Buddenbrooks," "Königliche Hoheit" und "Der Zauberberg,"* 3d rev. ed. (Bonn: H. Bouvier, 1980), as well as his *Der klassisch-moderne Roman in Deutschland: Thomas Mann, Alfred Döblin, Hermann Broch* (Stuttgart: W. Kohlhammer, 1983).

13. Hans Wysling, "Der Zauberberg," in *Thomas-Mann-Handbuch*, 414. On the Nietzschean influence, see Roger Archibal Nicholls, *Nietzsche in the Early Work of Thomas Mann* (Berkeley: University of California Press, 1955). See also T. J. Reed, *Thomas Mann: The Uses of Tradition* (Oxford: Clarendon, 1974), in which the notion of irony is foregrounded, as is the Nietzschean influence. On Schopenhauer's influence on the narrative form of the novel, see Børge Kristiansen, *Thomas Manns Zauberberg und Schopenhauers Metaphysik*, 2d ed. (Bonn: Bouvier, 1986). On patterns of musical composition, see Eckhard Heftrich, *Zauberbergmusik: Über Thomas Mann* (Frankfurt am Main: Klostermann, 1975); Agnes Schlee, *Wandlungen musikalischer Strukturen in Werke Thomas Manns: Vom Leitmotiv zur Zwölftonreihe* (Frankfurt am Main: Lang, 1981); and Ute Jung's brief study, *Die Musikphilosophie Thomas Manns* (Regensburg: Gustav Bosse, 1969).

14. Friedrich Nietzsche, *The Birth of Tragedy*, in *Basic Writings of Nietzsche*, trans. Walter Kaufmann (New York: Modern Library, 1968), 42.

15. There is more to the mountain motif. In the early twentieth century, as Siegfried Kracauer has observed, the mountain was a celebrated topic in German film and literature. It was popular, particularly among students and intellectuals, to climb up a mountain so as to gain perspective on the world below: "Full of Promethean promptings, they would climb up some dangerous 'chimney,' then quietly smoke their pipes on the summit, and with infinite pride look down on what they called 'valley pigs'— those plebeian crowds who never made an effort to elevate themselves to lofty heights" (*From Caligari to Hitler: A Psychological History of the German Film* [Princeton, N.J.: Princeton University Press, 1947], 111).

16. In an influential study that paved the way for the reception of *The Magic Mountain*, and that Mann himself endorsed, Hermann J. Weigand argues that Mann's novel is inscribed in the German romanticist tradition, both from a literary and a philosophical point of view. One of the earliest comprehensive studies of Mann's authorship thus is written by an American scholar and also focuses on *The Magic Mountain*. The fact that Mann endorsed it has no doubt contributed to its status. See Hermann J. Weigand, *The Magic Mountain: A Study of Thomas Mann's Novel "Der Zauberberg"* (Chapel Hill: University of North Carolina Press, 1964); first published as *Thomas Mann's Novel "Der Zauberberg"* (New York: Appleton, 1933).

17. Ernest Newman, *The Unconscious Beethoven: An Essay in Musical Psychology*, rev. ed. (New York: A. A. Knopf, 1930), 46. Quoted in Gunilla Bergsten, *Thomas Manns Doktor Faustus: Untersuchungen zu den Quellen und zur Struktur des Romans* (Stockholm: Scandinavian University Books, 1963), 84. Bergsten expounds on the common idea that disease is intimately linked to artistic genius (68–90). See also Susan Sontag, *Illness as Metaphor* (New York: Farrar, Straus and Giroux, 1978).

18. See, for example, Lotti Sandt, *Mythos und Symbolik im Zauberberg von Thomas Mann* (Bern: P. Haupt, 1979). See also Wysling, "Der Zauberberg," esp. 401–8.

19. For a discussion of *Der Zauberberg* as a "time-novel" influenced by Bergson, see, for example, Brennan, *Three Philosophical Novelists: James Joyce, André Gide, Thomas Mann* (New York: Macmillan, 1964), 143–51. On the implicit theories of time in the novel, see Francis Bulhof, *Transpersonalismus und Synchronizität: Wiederholung als Strukturelement in Thomas Manns "Zauberberg"* (Groningen: Drukkerij van Denderen, 1966), 112–47.

20. On the progressive role of visual technology in Bauhaus culture, see László Moholy-Nagy's classic treatise, *Painting, Photography, Film*, with a note by Hans M. Wingler and postscript by Otto Stelzer, trans. Janet Seligman (Cambridge, Mass.: MIT Press, 1969), quotation, 69. On the ideological repercussions of the German modernization process in the early twentieth century, particularly as they pertain to conservative attempts to reconcile technology with *Kultur*, see Jeffrey Herf, *Reactionary Modernism: Technology, Culture, and Politics in Weimar and the Third Reich* (Cambridge: Cambridge University Press, 1984).

21. Franco Moretti, *The Way of the World: The Bildungsroman in European Culture* (London: Verso, 1987), 55.

22. Moretti, *Way of the World*, 63.

23. Erich Heller, among others, subscribes to this view: "The sanatorium is Europe. It is also the world. Man is the patient" (*The Ironic German: A Study of Thomas Mann* [Boston: Little, Brown, 1958], 15).

24. Oliver Wendell Holmes, "The Stereoscope and the Stereograph," *Atlantic Monthly* (1859); reprinted in *Classic Essays on Photography*, ed. Alan Trachtenberg (New Haven, Conn.: Leete's Island Books, 1980), 82.

25. This argument, however, is not fully elaborated by Crary. For a wide-ranging study of the emergence of the European so-called science of labor and its concomitant notion of the human body as a machine, see Anson Rabinbach, *The Human Motor: Energy, Fatigue, and the Origins of Modernity* (New York: Basic Books, 1990).

26. Jay, *Downcast Eyes*, 187.

27. Gerhard Härle, in a psychoanalytic study of *The Magic Mountain* and its homoerotic thematics, has discussed Mann's narrative technique in terms of *Doppeloptik*, a notion derived from Wagnerian aesthetics. Härle reflects upon the gaze of desire in metaphorical terms, but fails to trace its dynamics in the novel itself, save for the observation that Castorp is drawn to Chauchat because of her "Pribislav" eyes; see Härle, *Die Gestalt des Schönen: Untersuchung zur Homosexualitätsthematik in Thomas Manns Roman "Der Zauberberg"* (Königstein: Hain, 1986). On the role of Chauchat's gaze, see Iso Camartin, "Die Augenlehre der Madame Chauchat," in *Die Bibliothek von Pila* (Frankfurt am Main: Suhrkamp, 1994), 162–82.

28. For a comprehensive discussion of Sander's photographic enterprise, including its historical context, see Ulrich Keller's introduction to August Sander, *Citizens of the Twentieth Century: Portrait Photographs, 1892–1952*, ed. Günther Sander, trans. Linda Keller (Cambridge, Mass.: MIT Press, 1986), 1–62.

29. Keller, introduction, 40.

30. Ernst Bloch, *Heritage of Our Times*, trans. Neville Plaice and Stephen Plaice (Berkeley: University of California Press, 1991), 106. "Germany in general," Bloch writes, "which had managed no bourgeois revolution up to 1918, is—unlike England,

and especially France—the classical land of non-contemporaneity, i.e. of unsurmounted remnants of older economic being and consciousness" (106).

31. On this encounter, see also Kristiansen, *Thomas Manns Zauberberg und Schopenhauers Metaphysik*, 98–103. In his view, however, Castorp's ambivalence testifies to his metaphysical *Steigerung*, a process marked by a Schopenhauerian dialectics of *Form* and *Unform*. The encounter with Settembrini accordingly represents an "Auflösung der 'flachländischen' Formwelt" on Castorp's part, *Formwelt* here understood as a predominantly metaphysical category rather than a social one.

32. Although Castorp quickly corrects his reading of Settembrini's social status, the image of the organ grinder persists. Every time the protagonist is overcome by a need to belittle his powerful mentor, the image emerges again, as in Castorp's dream (*MM* 157/*Z* 220); or, after an intense discussion, when Castorp attempts to write a letter to his relatives: "By now, he no longer felt like taking up the task of writing. The organ-grinder and his insinuations had definitely spoiled the mood for it" (*MM* 220/*Z* 308); or when he dwells on Clavdia's X-ray plate: "And in considering that inner aspect, he also thought of Settembrini, the pedagogic organ-grinder" (*MM* 383/*Z* 533).

33. On the "geography of Wilhelmine Culture" and the discourse of *Kultur* versus *Zivilisation*, see Russell Berman, *The Rise of the Modern German Novel: Crisis and Charisma* (Cambridge, Mass.: Harvard University Press, 1986), 1–54.

34. Norbert Elias, *The History of Manners*, vol. 1 of *The Civilizing Process*, trans. Edmund Jephcott (New York: Pantheon, 1982), 3–50.

35. Elias, *History of Manners*, 8.

36. See Benedict Anderson, *Imagined Communities: Reflections on the Origin and Spread of Nationalism*, 2d ed. (London: Verso, 1991), esp. 1–7.

37. Georg Lukács, "In Search of Bourgeois Man," in *Essays on Thomas Mann*, trans. Stanley Mitchell (New York: Grosset and Dunlap, 1965), 13–46.

38. In his brief discussion of *The Magic Mountain*, Peter Brooks usefully approaches Mann's narrative as a "Modernist discussion of the body," foregrounding the discovery of the interior; see Brooks, *Body Work: Objects of Desire in Modern Narrative* (Cambridge, Mass.: Harvard University Press, 1993), 263–66.

39. The sanatorium is remotely affiliated with those institutions whose disciplinary practices Foucault subjects to theorization—the prison, the asylum, and other restricted spaces with thoroughly regulated regimens directed at the body by way of the soul. In *Discipline and Punish: The Birth of the Prison*, trans. Alan Sheridan (New York: Pantheon, 1977), Foucault proposes that identity formation builds on the transforming internalization of the other's gaze or norms. In this sense, Castorp's process of *Bildung* is a disciplinary practice, a "technology of the self." The difference is that it is not enforced. The notion of the technology of the self would therefore have to be supplemented with the notion of the "care of the self," Foucault's designation for how, in ancient Greece, the free male fashioned his selfhood by means of subjecting himself to carefully regulated bodily practices—diet, exercise, sex, and so on. See Foucault, *The Use of Pleasure*, vol. 2 of *The History of Sexuality*, trans. Robert Hurley (New York: Pantheon, 1985); and *The Care of the Self*, vol. 3 of *The History of Sexuality*, trans. Robert Hurley (New York: Pantheon, 1986).

40. On the history of thermometry, see Stanley Joel Reiser, *Medicine and the Reign of Technology* (Cambridge: Cambridge University Press, 1978), 110–21.

41. No doubt Mann renders Dr Behrens as a quack doctor, in particular since the "moist spot" on Castorp's lungs will prove to be a false diagnosis. Interestingly enough, *The Magic Mountain* sparked a debate in the medical community in Germany. See Heinz Sauereßig, "Die medizinische Region des 'Zauberberg,' " in *Die Entstehung des Romans "Der Zauberberg,"* ed. Heinz Sauereßig (Biberach an der Riss: Wege und Gestalten, 1965), 25–34.

42. Michel Foucault, *The Birth of the Clinic: An Archaeology of Medical Perception,* trans. A. M. Sheridan Smith (New York: Pantheon, 1973). See also Reiser, *Medicine,* 1–22.

43. Foucault, *Birth of the Clinic,* 164. This argument is then confirmed by Reiser's etymological observation with regard to the stethoscope, the first medical instrument used to detect and study sounds produced in the body, developed in the early nineteenth century. The designation derives from the Greek words for "chest" and "I view." Furthermore, Reiser makes a valuable observation concerning the language of medicine: "One metaphor that recurred regularly in the medical literature between 1820 and 1850 was 'seeing' disease by listening through the stethoscope: 'We anatomize by auscultation (if I may say so), while the patient is yet alive', proclaimed a doctor, for whom the ear became an eye through auscultation. It is a 'window in the breast through which we can see the precise state of things within,' insisted another" (*Medicine,* 30).

44. Foucault, *Birth,* 195. Translation amended.

45. Bruno Latour, *Science in Action: How to Follow Scientists and Engineers through Society* (Cambridge, Mass.: Harvard University Press, 1987), esp. 138–40.

46. Ibid.

47. See Otto Glasser's classic account, *Wilhelm Conrad Röntgen and the Early History of the Roentgen Rays* (Springfield: C. C. Thomas, 1934), 1–28.

48. See, for example, Röntgen entries in *Asimov's Biographical Encyclopedia of Science and Technology,* 2d ed. (Garden City, N.Y.: Doubleday, 1982); *The History of Science and Technology: A Narrative Chronology* (New York: Facts on File, 1988); and Frank N. Magill, ed., *Great Events from History II: Science and Technology* (Pasadena, Calif.: Salem, 1991).

49. Glasser, *Wilhelm Conrad Röntgen,* 32.

50. Lisa Cartwright conceives of the effects of X-ray technology on the notion of the human body in terms of decomposition, usefully connecting this perceptual matrix to the new optical spaces and different modes of representing the human body that cinematography makes available. See Cartwright, *Screening the Body: Tracing Medicine's Visual Culture* (Minneapolis: University of Minnesota Press, 1995), 107–42. Commenting on *The Magic Mountain,* she suggests that the novel indicates that it is impossible to consider "medical techniques apart from their cultural meanings vis-à-vis life, sexuality, health, and death" (123).

51. Reiser, *Medicine,* 62, 60.

52. Ibid., 61. For reactions in the popular press, see Glasser, *Wilhelm Conrad Röntgen,* 199–209.

53. Leo Marx, who first used the term, has observed that during the nineteenth century, the American rhetoric of progress often adopted the rhetoric of the technological sublime; see Marx, *The Machine in the Garden: Technology and the Pastoral Ideal in America* (New York: Oxford University Press, 1964), esp. 195–207. See also Rosalind H.

Williams's discussion of European underground explorations as a source of images of the technological sublime, *Notes on the Underground: An Essay on Technology, Society, and the Imagination* (Cambridge, Mass.: MIT Press, 1990), 82–120.

54. On the motif of the hand in *The Magic Mountain*, see Erkme Joseph, "Das Motiv 'Hand'—ein Beispiel für Leitmotivtechnik im *Zauberberg* von Thomas Mann," in *Literaturwissenschaftliches Jahrbuch*, ed. Theodor Berchem et al. (Berlin: Duncker und Humblot, 1993), 131–69.

55. Weigand has also underscored the centrality of the X-ray episode for Castorp's quest for knowledge; see Weigand, *The Magic Mountain*, 23.

56. Sigmund Freud, *Notes upon a Case of Obsessional Neurosis*, in vol. 10 of *The Standard Edition of the Complete Psychological Writings of Sigmund Freud* (London: Hogarth Press, 1955), 158–249. On Freud's notion of epistemophilia, see Toril Moi, "Patriarchal Thought and the Drive for Knowledge," in *Between Feminism and Psychoanalysis*, ed. Teresa Brennan (London: Routledge, 1989); reprinted in Toril Moi, *What Is a Woman? And Other Essays* (Oxford: Oxford University Press, 1999), esp. 361–68.

57. In 1928, Thomas Mann wrote an essay on film, explaining that while he often goes to the cinema and derives pleasure from watching film, he does not consider cinema an art form: "I was speaking of a 'phenomenon of life,' because I believe that film—you will excuse me—has little to do with *art*; and it would be a mistake to approach film using criteria taken from the world of art. [. . .] Film is not art, it is life and reality; and its effects, in all its mobile silence, are crudely sensational in comparison to the spiritual effects of art" (Mann, "Über den Film," in *Kino-Debatte: Texte zum Verhältnis von Literatur und Film, 1909–1929*, ed. Anton Kaes [Tübingen: Niemeyer, 1978], 164). In a 1928 interview, he affirmed his views on cinema: "Film is an art surrogate, not art. It has very little to do with art, because art is cold, in the supreme sense of the word: mediated, not crude, not sensational—film, on the other hand, is an immediate reproduction of life" (Victor Wittner, interview with Thomas Mann, *Neue Freie Presse* [Wien], 28 October 1928; reprinted in *Frage und Antwort: Interviews mit Thomas Mann, 1909–1955*, ed. Volkmar Hansen and Gert Heine [Hamburg: Knaus, 1983], 136). On German intellectuals and film, see Anton Kaes's excellent introduction to *Kino-Debatte*, 1–36.

58. Jürgen Kolbe and Christoph Schmidt, respectively, have argued that the episode alludes to a specific film, *Sumurun*, by Ernst Lubitsch. Mann first saw Lubitsch's Orientalist film in 1920, at the Lichtspieltheater am Sendlingerthor in Munich, and then recorded the visit in his diary. See Jürgen Kolbe, *Heller Zauber: Thomas Mann in München, 1894–1933* (Berlin: Siedler, 1987), 379–81. Christoph Schmidt has detected Spenglerian overtones in the cinema episode, arguing that the spirit of the passage overlaps with a recognizably conservative stance toward the new mass medium. See Schmidt, " 'Gejagte Vorgänge voll Pracht und Nacktheit': Eine unbekannte kinematographische Quelle zu Thomas Manns Roman 'Der Zauberberg,' " *Wirkendes Wort* 38, no. 1 (March–April 1988): 1–5. See also Thomas Sprecher, *Davos im Zauberberg: Thomas Manns Roman und sein Schauplatz* (Munich: W. Fink, 1996), 222.

59. On the technologization of the production and reception of music, see Michael Chanan, *Repeated Takes: A Short History of Recording and Its Effects on Music* (London: Verso, 1995). Chanan dwells briefly on the gramophone episode in Mann's novel (41–43).

60. Brennan has discussed the remarkable fact that Wagner's music is not among the records Castorp listens to; and yet Wagner's formal presence in the novel itself is

striking: "If no explicit reference to the music of Wagner is made in *The Magic Mountain,* a methodological device characteristic of the composer is definitely employed there. It is the *Leitmotiv*" (*Thomas Mann's World* [New York: Columbia University Press, 1942], 99).

61. On Nietzsche and music, see Curt Paul Janz, "The Form-Content Problem in Nietzsche's Conception of Music," as well as Michael Allen Gillespie, "Nietzsche's Musical Politics," in *Nietzsche's New Seas: Explorations in Philosophy, Aesthetics, and Politics,* ed. Michael Allen Gillespie and Tracy B. Strong (Chicago: University of Chicago Press, 1988), 97–116; 117–49.

62. Ernst Robert Curtius, "Thomas Manns 'Zauberberg,' " *Luxemburger Zeitung,* 9 January 1925; reprinted in Heinz Sauereßig, ed., *Die Entstehung des Romans "Der Zauberberg,"* 55. In the same vein, Curtius's general conclusion is fraught with allusions to a German geopolitics of culture: "A peculiar transformation has taken place: what in the French or English or Russian novel would be psychological here appears as dialectical and metaphysical. This is one of the eminently German traits of Thomas Mann's work. French literature is psychological; German literature, in its innermost nature, is metaphysical. *All great German works are Faustian. Reflective in a Faustian fashion, experimental in a magical manner, crafted in a musical way—such is also Thomas Mann's work*" (53; emphasis in the original). For a contrasting perspective on music and the German character, see Hermann Hesse's 1927 novel *Steppenwolf,* trans. Basil Creighton, revised by Joseph Mileck and Horst Frenz (New York: Henry Holt, 1990), 135–36: "For long during this night's walk I had reflected upon the significance of my relation to music, and not for the first time recognized this appealing and fatal relation as the destiny of the entire German spirit. In the German spirit the matriarchal link with nature rules in the form of the hegemony of music to an extent unknown in any other people. We intellectuals, instead of fighting against this tendency like men, and rendering obedience to the spirit, the Logos, the Word, and gaining a hearing for it, are all dreaming of a speech without words that utters the inexpressible and gives form to the formless. [. . .] And so the German spirit, carousing in music, [. . .] has left the greater parts of its practical gifts to decay."

63. Hörisch, " 'Die deutsche Seele up to date,' " 23.

64. One of the few critics who have commented on the role of machinery in the novel is Weigand, but these observations are immediately subsumed under his major concern, which is to show the great extent to which Mann's narrative is immersed in the German romanticist tradition; see Weigand, *The Magic Mountain,* 81, 158.

65. Foster, *Compulsive Beauty,* 148.

66. Nietzsche, *Human, All Too Human: A Book for Free Spirits,* trans. R. J. Hollingdale (Cambridge: Cambridge University Press, 1986), 100. Nietzsche develops this line of thought in his later works on Wagner, where he argues that the composer had succumbed to a logic of means and ends.

67. Theodor W. Adorno, *In Search of Wagner,* trans. Rodney Livingstone (London: Verso, 1991), 102.

68. Thomas Mann, preface to the American edition of *Stories of Three Decades,* trans. H. T. Lowe-Porter (New York: A. A. Knopf, 1936), vi.

69. Sauereßig, "Die medizinische Region des 'Zauberberg,' " 25.

70. I draw here on Michèle Le Doeuff's notion of the philosophical imaginary. As

Le Doeuff argues in *The Philosophical Imaginary*, trans. Colin Gordon (Stanford, Calif.: Stanford University Press, 1989), when philosophers such as Descartes, Galileo, Rousseau, and Kant have recourse to images, even when the image in question belongs in the philosophical tradition, the effect is frequently an ambiguous and contradictory discourse, often revealing a nostalgia for the power that philosophy once had.

3. The Education of the Senses

1. In preparing this chapter, I have consulted the following bibliographies and works of reference: Douglas W. Alden, *Marcel Proust and His Critics* (Los Angeles: Lymanhouse, 1940); Douglas W. Alden and Richard Brooks, eds., *The Twentieth Century*, vol. 6, part 1 of *A Critical Bibliography of French Literature*, ed. David Clark Cabeen (Syracuse, N.Y.: Syracuse University Press, 1980); Victor E. Graham, *Bibliographie des études sur Marcel Proust et son œuvre* (Geneva: Droz, 1976); Janet C. Stock, *Marcel Proust: A Reference Guide, 1950–1970* (Boston: G. K. Hall, 1991); *Bibliographie der französischen Literaturwissenschaft* (Frankfurt am Main: Klostermann, 1970–); Elisabeth Russell Taylor, *Marcel Proust and His Contexts: A Critical Bibliography of English-Language Scholarship* (New York: Garland, 1981); Sigbrit Swahn, *Proust dans la recherche littéraire: problèmes, méthodes, approches nouvelles* (Lund: Liber, 1979); Leighton Hodson, ed., *Marcel Proust: The Critical Heritage* (London: Routledge, 1989); Jean-Yves Tadié, ed., *Lectures de Proust* (Paris: Armand Colin, 1971), as well as his "Bilan critique," in *Proust* (Paris: P. Belfond, 1983), 155–231; Joyce N. Megay, "Proust et la Nouvelle Critique," *French Review* 47, no. 6 (spring 1974): 120–28.

2. On the relation between railway travel and the emergence of new clinical diagnoses, see Wolfgang Schivelbusch, *The Railway Journey: The Industrialization of Time and Space in the Nineteenth Century* (Berkeley: University of California Press, 1986), 113–58.

3. The cinematic train is another matter. When the Lumière brothers first screened *L'arrivé d'un train en gare* in 1895, the train elicited a new kind of perceptual agony, bespeaking the not-yet-domesticated visual habits of the spectators, for audiences reportedly stampeded when the powerful locomotive came rushing toward them. See David Cook, *A History of Narrative Film*, 2d ed. (New York: Norton, 1990), 11. In like spirit, Maxim Gorky, having attended one of the earliest screenings of the Lumière program in Russia, reported that "suddenly something clicks, everything vanishes and a train appears on the screen. It speeds straight at you—watch out! It seems as though it will plunge into the darkness in which you sit, turning you into a ripped sack full of lacerated flesh and splintered bones" (Gorky, review of the Lumière program, in *Kino: A History of the Russian and Soviet Film*, ed. Jay Leyda, 3d ed. [Princeton, N.J.: Princeton University Press, 1983], 408).

4. The English translation is based on the 1954 edition of *A la recherche*. In the new French edition, edited by Jean-Yves Tadié et al., it is primarily the three volumes of *A la recherche* that were published posthumously that have been subject to revision—*La prisonnière* (1923), *Albertine disparue* (1925), and *Le temps retrouvé* (1927).

5. According to Jean-Yves Tadié, Proust started work on his great novel toward the end of 1908; see Tadié, *Proust: Biographie* (Paris: Gallimard, 1996), esp. 619–31.

6. For a study of the role of vision in Proust's novel, with a focus on optical metaphors, see Roger Shattuck's pioneering work, *Proust's Binoculars: A Study of*

Memory, Time, and Recognition in "A la recherche du temps perdu" (New York: Random House, 1963); quotation, 6; emphasis in the original. See also Shattuck, *Proust's Way: A Field Guide to "In Search of Lost Time"* (New York: Norton, 2000), 99–136.

7. On the proto-cinematic nature of the magic lantern and its role in *Remembrance*, see Marie-Claire Ropars-Wuilleumier, "L'image-mémoire, ou l'écriture de l'oubli," *Hors cadre* 9 (1991): 117–33. Ropars-Wuilleumier even suggests that the magic lantern offers the model on which processes of memory are staged in *Remembrance*.

8. On Edison's development of the kinetoscope and its affinities with Marey's apparatuses, see Marta Braun, *Picturing Time: The Work of Etienne-Jules Marey* (Chicago: University of Chicago Press, 1992), 187–93.

9. William C. Carter, *The Proustian Quest* (New York: New York University Press, 1992), 11.

10. For the account of Marey's scientific enterprise below I am indebted to Marta Braun's path-breaking study, *Picturing Time*. I also found the references to Bergson, Souriau, and Sizeranne in Braun. On the significance of Marey's work, with a particular focus on the emerging science of labour, see Anson Rabinbach's important study, *The Human Motor: Energy, Fatigue, and the Origins of Social Modernity* (New York: Basic Books, 1990). See also François Dagognet, *Etienne-Jules Marey: A Passion for the Trace*, trans. Robert Galeta, with Jeanine Herman (New York: Zone, 1992); and Siegfried Giedion, *Mechanization Takes Command: A Contribution to Anonymous History* [1948] (New York: Norton, 1969), 17–30. On Marey's impact on medicine and diagnostic practices, see Stanley Joel Reiser, *Medicine and the Reign of Technology* (Cambridge: Cambridge University Press, 1978). For an analysis of how Marey conceptualized time as a problem of representation, see Mary Ann Doane, "Temporality, Storage, Legibility: Freud, Marey, and the Cinema," *Critical Inquiry* 22, no. 2 (winter 1996): 313–43.

11. Etienne-Jules Marey, *La méthode graphique dans les sciences expérimentales et principalement en physiologie et en médecine* (Paris: G. Masson, 1878), i.

12. Ibid., 108.

13. Friedrich A. Kittler, *Discourse Networks, 1800/1900*, trans. Michael Metteer, with Chris Cullens, with a foreword by David E. Wellbery (Stanford, Calif.: Stanford University Press, 1990), 230.

14. Bruno Latour, *Science in Action: How to Follow Scientists and Engineers through Society* (Cambridge, Mass.: Harvard University Press, 1987), 230.

15. Giedion, *Mechanization Takes Command*, 24.

16. Along with a vast amount of unknown bronze sculptures by Degas, the levitating horse was found after the artist's death. As for the date of conception, only one thing is certain: it cannot have been conceived before 1878, when Muybridge's photographic experiments began spreading over the world. Degas's sculpture is in the Metropolitan Museum in New York.

17. Braun, *Picturing Time*, 264.

18. Pierre Cabanne, *Dialogues with Marcel Duchamp*, trans. Ron Padgett (New York: Da Capo Press, 1987), 28.

19. Robert de la Sizeranne, *La photographie, est-elle un art?* (Paris: Hachette, 1899), 42–43. Translated in Braun, *Picturing Time*, 275; emphasis in the original.

20. Braun, *Picturing Time*, 278–81; Rabinbach, *Human Motor*, 110–14.

21. For a detailed account of how cinematography emerged out of chronophotography, see Braun, *Picturing Time*, 150–98.

22. Bergson, *Creative Evolution*, trans. Arthur Mitchell (New York: Random House, 1944), 361.

23. Ibid.

24. On these and related debates, see Paul Virilio, *The Vision Machine*, trans. Julie Rose (Bloomington: Indiana University Press, 1994), 1–18.

25. See Julia Kristeva, *Time and Sense: Proust and the Experience of Literature*, trans. Ross Guberman (New York: Columbia University Press, 1996), 392, n. 4.

26. J. Theodore Johnson, Jr., has suggested that Elstir's studio has the appearance of a giant magic lantern, and that the artist personifies the light source and projects the images. As Johnson notes, Proust himself makes this connection elsewhere in the novel; see Johnson, "La lanterne magique: Proust's Metaphorical Toy," *L'esprit créateur* 11, no. 1 (1971): 28–30. Roxanne Hanney has shown that the dark room is a recurrent thematic element in Proust; the narrator's consciousness, for example, is often likened to a dark room, *chambre noire*; see Hanney, "Proust and Negative Plates: Photography and the Photographic Process in *A la recherche du temps perdu*," *Romantic Review* 74, no. 3 (1983): 342–54.

27. Anne Henry, "Quand une peinture métaphysique sert de propédeutique à l'écriture: Les métaphores d'Elstir dans *A la recherche du temps perdu*," in *La critique artistique: un genre littéraire*, ed. Jean Gaulmier (Paris: Presses Universitaires de France, 1983), 224.

28. Nathalie Sarraute, *The Age of Suspicion: Essays on the Novel*, trans. Maria Jolas (New York: George Braziller, 1963), 72.

29. *The Works of John Ruskin*, ed. E. T. Cook and Alexander Wedderburn, 42 vols. (London: George Allen, 1903–12), 33:210. Hereafter referred to as *WJR*, with page references given parenthetically in the text.

30. Marcel Proust, "Preface to *La bible d'Amiens*," in *On Reading Ruskin: Prefaces to "La bible d'Amiens" and "Sésame et les lys," with Selections from the Notes to the Translated Texts*, trans. and ed. Jean Autret, William Burford, and Phillip J. Wolfe, with an introduction by Richard Macksey (New Haven, Conn.: Yale University Press, 1987), 41–42.

31. See Jean Autret, *L'influence de Ruskin sur la vie, les idées et l'œuvre de Marcel Proust* (Geneva: Droz, 1955). On Proust's interest in Ruskin, see also Richard Macksey, introduction to Proust, *On Reading Ruskin*, xiii–liii. For a critical discussion, see Anne Henry, *Marcel Proust: Théories pour une esthétique* (Paris: Klincksieck, 1971), "Contre Ruskin," 166–257.

32. The visual arts play an important role in Proust's novel. On the influence of premodernist art, notably Renoir, Turner, and Chardin, see Juliette Monnin-Hornung's classic study, *Proust et la peinture* (Geneva: Droz, 1951). On the great extent to which Proust relied on specific paintings for his descriptions, see John M. Cocking, "Proust and Painting," in *Proust: Collected Essays on the Writer and His Art* (Cambridge: Cambridge University Press, 1982). See also Johnson, "Proust and Painting," in *Critical Essays on Marcel Proust*, ed. Barbara Bucknall (Boston: G. K. Hall, 1987). For studies that argue the essentially modernist character of Proust's aesthetic enterprise, see Reinhold Hohl, "Marcel Proust in neuer Sicht: Kubismus und Futurismus in seinem Romanwerk," *Neue Rundschau* 88 (1977): 54–72; Theodore J. Johnson, Jr.,

"Proust's 'Impressionism' Reconsidered in the Light of the Visual Arts of the Twenti-eth Century," in *Twentieth Century French Fiction*, ed. George Stambolian (New Brunswick, N.J.: Rutgers University Press, 1975), 27–56; Diane R. Leonard, "Literary Evolution and the Principle of Perceptibility: The Case of Ruskin, Proust, and Mod-ernism," in *Proceedings of the Tenth Congress of the International Comparative Litera-ture Association*, vol. 1, ed. Anna Balakian et al. (New York: Garland, 1985), 132–37.

33. Commenting on Proust's railroad passage, Keith Cohen similarly observes that "the constant mobility of the point of observation contributes significantly to the total effect. [. . .] This description [of the sunrise] is really just an extreme case of what happens throughout the *Recherche*. No object, thought, or person can be grasped di-rectly in its totality, but must rather be apprehended successively, from various frag-mented viewpoints" (*Film and Fiction: The Dynamics of Exchange* [New Haven, Conn.: Yale University Press, 1979], 164–65).

34. Schivelbusch, *Railway Journey*, 64.

35. As I discuss below, this is the major argument in Lynne Kirby's *Parallel Tracks: The Railroad and Silent Cinema* (Durham, N.C.: Duke University Press, 1997).

36. The snapshot of Saint-Loup is a brilliant portrait, the image of the fluttering butterfly prefiguring the narrator's futile and constantly suspended attempts to deci-pher the "true" nature of the marquis—in the history of art the butterfly signifies *psy-che*, the human soul. The feature that more than any other propels the narrator's scopophilic will-to-know is Saint-Loup's homosexuality. Throughout the novel, the narrator tries to frame Saint-Loup photographically. Put differently, Proust uses pho-tographic modes of representation in order to visualize the difficulties inherent in the narrator's pursuit of secure knowledge. On the epistemology of the photograph and the question of homosexuality in Proust, see Mieke Bal, *The Mottled Screen: Reading Proust Visually* (Stanford, Calif.: Stanford University Press, 1998), 226–30.

37. Macksey, introduction, xviii.

38. Ibid., xix.

39. Proust, "Pèlerinages ruskiniens en France," in *CSB*, 441.

40. See Proust, "Preface to *La bible d'Amiens*," in *On Reading Ruskin*, 43.

41. On the picturesque and its significance for the early Ruskin, see Robert Hewi-son, *John Ruskin: The Argument of the Eye* (Princeton, N.J.: Princeton University Press, 1976), 30–53.

42. Ibid., 34–35.

43. *JWR* 3:169; see also 14:358–60; 15:353; and 21:236–37. Ruskin nevertheless recom-mended to his students of drawing the use of an equally mechanical means of repro-ducing nature, namely, photography. Photographs were to be copied as a form of ex-ercise and compared for their "authority" (*WJR* 15:100–1; see also 21:294–95). In fact, the young Ruskin—the Ruskin of the picturesque period—was enthusiastic about the advent of daguerreotypy, claiming that it beat his own talent for drawing. "It is a noble invention—say what they will of it," Ruskin wrote from Venice in 1845, after having procured a few daguerreotypes of the palaces he had tried to draw, "and any one who has worked and blundered and stammered as I have done for four days, and then sees the thing he has been trying to do so long in vain, done perfectly and faultlessly in half a minute, won't abuse it afterwards" (*WJR* 3:210n).

44. For other examples of Ruskin's contempt for railway travel, see *WJR* 5:380–81;

8:158–59. On Ruskin's attitude toward the railway and its inherent modes of perception, see Schivelbusch, *Railway Journey,* 52–64.

45. Victor Graham, *The Imagery of Proust* (Oxford: Basil Blackwell, 1966), 33.

46. Susan Sontag, *On Photography* [1977] (New York: Farrar, Straus and Giroux, 1989), 164. On Proust's relation to photography, with an emphasis on how photography features in the novel, see Jean-François Chevrier, *Proust et la photographie* (Paris: Editions de l'Etoile, 1982). See also Brassaï, *Marcel Proust sous l'emprise de la photographie* (Paris: Gallimard, 1997); Stephen C. Infantino, *Photographic Vision in Proust* (New York: P. Lang, 1991); Hanney, "Proust and Negative Plates"; Anna Giubertoni, "Fotografia e aura nella narrativa di Marcel Proust," *Rivista di letterature moderne e comparate* (Florence) 28 (1975): 12–27; Mieke Bal, "Identification et apprentissage de la compassion: Proust et la photographie," in *(En)jeux de la communication romanesque: Hommage à Françoise van Rossum-Guyon,* ed. Suzan van Dijk and Christa Stevens (Amsterdam: Rodopi, 1994), 241–56.

47. Proust nevertheless used photographic prints as an aid to memory or as a point of departure for many of his depictions of paintings, locations, and buildings; see Jacques Nathan, *Citations, références et allusions de Marcel Proust dans "A la recherche du temps perdu,"* rev. ed. (Paris: Nizet, 1969), 15–16.

48. Shattuck, *Proust's Binoculars,* 146.

49. Rosalind Williams, *Notes on the Underground: An Essay on Technology, Society, and the Imagination* (Cambridge, Mass.: MIT Press, 1990), 82–120. On the category of the sublime in late nineteenth-century representations of the city, see Christophe den Tandt, *The Urban Sublime in American Literary Naturalism* (Urbana: University of Illinois Press, 1998). See also Joseph Tabbi, *Postmodern Sublime: American Literature and Technology from Mailer to Cyberpunk* (Ithaca, N.Y.: Cornell University Press, 1995).

50. Franz Kafka, "The Aeroplanes at Brescia," in *Franz Kafka: A Biography,* by Max Brod, trans. G. Humphreys Roberts (London: Secker and Warburg, 1948), 171–9. This enthusiastic description of aviation forms an interesting contrast to Kafka's otherwise highly ambiguous relationship to modern technology; see Klaus Benesch, "Writing Machines: Technology and the Failures of Representation in the Works of Franz Kafka," in *Reading Matters: Narratives in the New Media Ecology,* ed. Joseph Tabbi and Michael Wutz (Ithaca, N.Y.: Cornell University Press, 1997), 76–95. See also Hanns Zischler, *Kafka geht ins Kino* (Reinbek bei Hamburg: Rowohlt, 1996).

51. As Dirk Hoeges has underscored, in few works of literature are technologies of speed so well integrated as in Proust's novel. Focusing on so many discursive passages in *Remembrance,* Hoeges discusses the way in which Proust theorizes the impact of the train, the airplane, and the automobile on notions of time and space; see Hoeges, "Paris, Balbec und um Balbec herum: Zeit, Raum und Technik in der *Recherche,"* in *Marcel Proust: Motiv und Verfahren,* Marcel Proust Gesellschaft, vol. 4 (Frankfurt am Main: Insel, 1986), 31–53. For a study of the airplane motif in Proust's notebooks, see Karl Hölz, "Das Motiv des 'aéroplane' bei Marcel Proust: Die narrative Verwandlung eines Avant-Textes in einen Kon-text," in *Marcel Proust: Schreiben ohne Ende,* ed. Rainer Warning, Marcel Proust Gesellschaft, vol. 7 (Munich: Insel, 1994), 129–47. For an inventory of modern means of transport in Proust, see Roger Kempf, "Sur quelques véhicules," *L'Arc,* no. 47 (1971): 47–57. See also Eugène Nicole, "Les inventions modernes dans *La recherche du temps perdu,"* *Bulletin de la Société des Amis de Marcel Proust et des Amis de Combray* 36 (1986): 528–42.

52. Deleuze, *Proust and Signs: The Complete Text*, trans. Richard Howard (Minneapolis: University of Minnesota Press, 2000), 3–25.

53. Carter's important study *The Proustian Quest* explores the theme of speed both from a biographical and a textual view, esp. 1–22, 63–91. See also Wolfram Nitsch, "Phantasmen aus Benzin: Prousts Automobile in textgeschichtlicher Sicht," in *Marcel Proust: Schreiben ohne Ende*, 93–108. Ernst Robert Curtius is, to my knowledge, the first critic to have commented upon the importance of speed and motion in Proust's oeuvre; see Curtius, "Marcel Proust," in *Französischer Geist im Neuen Europa* (Stuttgart: Deutsche Verlags-Anstalt, 1925); reprinted as *Marcel Proust* (Berlin: Suhrkamp, 1952), 38–39; 83; 107–9.

54. Proust, *Correspondance*, ed. Philip Kolb, 21 vols. (Paris: Plon, 1970–93), 7:296; hereafter cited as *C*.

55. On the emerging automobile culture in the early twentieth century, see Pär Bergman, *"Modernolatria" et "Simultaneità": Recherches sur deux tendances dans l'avant-garde littéraire en Italie et en France à la veille de la première guerre mondiale* (Uppsala: Scandinavian University Books, 1962), 16–19. On the artistic interest in speed, see Stephen Kern, *The Culture of Time and Space, 1880–1918* (Cambridge, Mass.: Harvard University Press, 1983), 109–30.

56. Maurice Maeterlinck, "In an Automobile," trans. Alfred Sutro, in *The Double Garden*, trans. Alexander Teixeira de Mattos (New York: Dodd, Mead, 1904), 172; "En automobile," in *Le double jardin* (Paris: Bibliothèque Charpentier, 1904), 52. Page references, cited parenthetically in the text, refer first to the English translation and then to the French original.

57. Eugène Demolder, *L'Espagne en auto: Impressions de voyage* (Paris: Mercure de France, 1906).

58. A selected translation of Octave Mirbeau's *La 628—E 8* (Paris: Fasquelle, 1908) appears in Pierre Bonnard, *Sketches of a Journey: Travels in an Early Motorcar* (London: Philip Wilson Publishers, 1989), with illustrations by Pierre Bonnard (translation not credited). I found the reference to Mirbeau's novel in Bergman's *"Modernolatria" et "Simultaneità."*

59. Tom Gunning, "An Unseen Energy Swallows Space: The Space in Early Film and Its Relation to American Avant-Garde Film," in *Film Before Griffith*, ed. John L. Fell (Berkeley: University of California Press, 1983), 355–66.

60. Raymond Fielding, "Hale's Tours: Ultrarealism in the Pre-1910 Motion Picture," in *Film Before Griffith*, 116–17.

61. My account of these cinematic fairground attractions builds upon Gunning, "An Unseen Energy Swallows Space"; Fielding, "Hale's Tours"; Georges Sadoul, *Histoire générale du cinéma*, vol. 2, *Les pionniers du cinéma, 1897–1909* (Paris: Editions Denoël, 1947), 100–116; Kenneth Macgowan, *Behind the Screen: The History and Techniques of the Motion Pictures* (New York: Delacorte Press, 1965), 465–68; and Vanessa R. Schwartz, *Spectacular Realities: Early Mass Culture in Fin-de-Siècle Paris* (Berkeley: University of California Press, 1998), 149–76.

62. For contemporary news coverage of the *maréorama* and similar attractions at the 1900 Expo, see also *Le Figaro*'s article on *le ballon cinéorama*, 9 April, 1; on *le trottoir roulant*, 18 April, 1; on *le maréorama*, 17 July, 3; and on *le Phono-Cinéma-Théâtre*, 6 September, 3.

63. See Kirby, *Parallel Tracks*.

64. "Lev Tolstoy," appendix in Jay Leyda, ed., *Kino*, 410.

65. F. T. Marinetti, "The Founding and Manifesto of Futurism," in *Let's Murder the Moonshine: Selected Writings*, ed. R. W. Flint, trans. R. W. Flint and Arthur A. Coppotelli, with a preface by Marjorie Perloff (Los Angeles: Sun and Moon Classics, 1991), 49.

66. Proust reprinted "Impressions de route en automobile" under the title "Journées en automobile" in *Pastiches et mélanges* (1919). My citations refer to "Journées en automobile," in *CSB*, 63–69.

67. Tadié, *Proust: Biographie*, 596.

68. See Nitsch, "Phantasmen aus Benzin," esp. 98–102.

69. It may well be that Proust, in writing "Impressions de route en automobile," sought to surpass Maeterlinck's "En automobile." Both stories are set in the countryside in Normandy; both stress that the motorcar enables modern humans to experience as many landscapes and spectacles in a day as would formerly have demanded a whole lifetime; both attempt to represent the visual impressions along the road. A great admirer of Maeterlinck's work, Proust repeatedly returned to the celebrated writer and playwright in his literary texts and especially in his letters. He is likely to have read Maeterlinck's "En automobile." The story was published in *Le double jardin*, which came out in 1904, and Proust refers to the volume that same year. Furthermore, as Philip Kolb has noted, in the so-called *Carnet de 1908*, a notebook containing notes toward *Contre Sainte-Beuve*, Proust alludes to Maeterlinck's motoring piece: "Odeur des automobiles en campagne. Maeterlinck a tort" (Proust, *Carnet de 1908*, ed. Philip Kolb, Cahiers Marcel Proust, no. 8 [Paris: Gallimard, 1976], 70). Kolb notes, too, that Proust also alludes to the piece when, in *The Captive*, the smell of petrol reminds the narrator of his automobile excursions with Albertine. In addition, Proust produced a pastiche of Maeterlinck's style; at the end, he touches upon the automobile theme and the excitement of speed ("L'Affaire Lemoine par Maeterlinck," in *CSB*, 197–201). If Proust indeed sought to outdo Maeterlinck's motoring piece, it is an impulse that befits what the Martinville episode signifies: the birth of a writer, untouched by the anxiety of influence. See also Nitsch, "Phantasmen aus Benzin," 102–4.

70. Here as well as in the following quote from "Journées en automobile," my translation is, as far as possible, based on the English translation of the corresponding passages in *Swann's Way*.

71. Curtius, *Marcel Proust*, 83.

72. Most commentators have read the Martinville episode in the interpretive terms proposed by the novel itself, concentrating on the question of literary creation and the mysteries of perspective. A few analyses deviate from this pattern. Arguing that in Proust travel is more miraculous than involuntary memory, Georges Poulet suggests that the travel motif is a structural pretext for bringing together unrelated objects in space. Moreover, contrary to memory, the phenomena that travel conjoins do not depend upon resemblances (Poulet, *Proustian Space*, trans. Elliott Coleman [Baltimore: Johns Hopkins University Press, 1977], 73–88). In Deleuze's view, the Martinville episode fuels what he perceives as the novel's global theme: the hero's apprenticeship in the art of reading signs (Deleuze, *Proust and Signs*, 3–5 et passim). For Jean-Pierre Richard, the Martinville steeples are a typically Proustian hermeneutic object, but also an allegory to be unpacked along psychoanalytic lines (Richard, *Proust et le monde sensible*

[Paris: Seuil, 1974], 155–67). According to Diane Leonard, the Martinville episode testifies to Proust's modernism, as it introduces perspectival illusions and the question of depth into Proust's novel (Leonard, "Proust and Virginia Woolf, Ruskin and Roger Fry: Modernist Visual Dynamics," *Comparative Literature Studies* 18, no. 3 [September 1981]: 333–43). Linda A. Gordon claims that the episode is "a model of the perspectival relationships between subject and object" (Gordon, "The Martinville Steeplechase: Charting the Course," *Style* 22, no. 3 [fall 1988]: 402–9). For Howard Moss, the steeples are premonitions, as they embody a slice of Marcel's hidden, writerly future (Moss, *The Magic Lantern of Marcel Proust* [New York: Macmillan, 1962], 87–103). For a stylistic commentary, see Luzius Keller, *Proust Lesen* (Frankfurt am Main: Suhrkamp, 1991), 229–38.

73. Cohen, *Film and Fiction*, 160.

74. As I point out earlier, Proust reprinted "Impressions de route en automobile" in *Pastiches et mélanges* (1919). He then added two footnotes, explaining among other things that he had recycled parts of the *Figaro* article in *A la recherche du temps perdu;* he also remarks that in a not-yet-published volume of *A la recherche*, the publication of the "fragment" about the steeples in *Le Figaro* would be the subject of almost an entire chapter. The Martinville episode is widely thought of as a crucial episode in Proust's novel, but most critics have nevertheless neglected its derivation from "Impressions de route en automobile." Curtius is an exception, as are Bernard Guyon and Wolfram Nitsch. Guyon seeks to show that the Martinville episode is a "poetical transposition" of the piece in *Le Figaro*. He also argues that the novel's genesis must be sought in the 1907 article; see Guyon, "Marcel Proust et le mystère de la création littéraire: Essai d'explication des 'Clochers de Martinville,' " *Annales de la Faculté des lettres de Toulouse*, no. 1–2 (1955): 37–65. Nitsch, for his part, discusses the genesis of the automobile motif in Proust at large, his principal aim being to study how the author's attitude toward the motif changes over the years. He concentrates on the *Figaro* article, usefully connecting the visual enigmas triggered by the car ride to Elstir's painterly aesthetic; see Nitsch, "Phantasmen aus Benzin," 93–108. Carter, although stopping short of textual analysis, has also foregrounded the connection between the *Figaro* article and the Martinville episode; see Carter, *Proustian Quest*, 135–37.

75. Guyon, "Marcel Proust et le mystère de la création littéraire," 58.

76. Think of the episode about the death of the writer Bergotte in *The Captive*, where Proust accomplishes one of his most moving and pregnant images: "They buried him, but all through that night of mourning, in the lighted shop-windows, his books, arranged three by three, kept vigil like angels with outspread wings and seemed, for him who was no more, the symbol of his resurrection" (*REM* 3:186/*RTP* 3:693).

77. As Proust's 1907 correspondence shows, he spent August and part of September in Cabourg, and went on trips "en automobile fermée" to architecturally significant sites in Normandy. Judging from the letters, "Impressions de route en automobile" appears to have received an enthusiastic response from his friends. See, for example, Proust's letter to Mme Straus (*C* 7:315–16), in which Proust also refers to Agostinelli's admiring letter. See also Daniel Halévy's letter to Proust, which, as Philip Kolb has suggested, was part of an exchange apropos of the motoring article (*C* 7:320–22).

78. *C* 7:260–61.

79. Autret, *L'influence de Ruskin*, 85.

80. Quoted in Gerald Silk, "The Automobile in Art," in *Automobile and Culture*, ed. Gerald Silk (New York: Abrams, 1984), 75. See also Silk's interesting comments on *Le Pare-brise* (75). Interestingly, Matisse executed a series of paintings of landscapes seen through windshields; see Jack Flam, *Matisse: The Man and His Art, 1869–1918* (Ithaca, N.Y.: Cornell University Press, 1986), 460–61, 468–69.

81. Fernand Léger, "Contemporary Achievements in Painting," in *Functions of Painting*, ed. Edward F. Fry, trans. Alexandra Anderson (New York: Viking, 1973), 11. Similarly, László Moholy-Nagy suggests that "motion, accelerated to high speed, changes the appearance of the objects and makes it impossible to grasp their details. There is a clearly recognizable difference between the visual experience of a pedestrian and a driver in viewing objects. The motor car driver or airplane pilot can bring distant and unrelated landmarks into spatial relationships unknown to the pedestrian" (*Vision in Motion* [Chicago: Paul Theobald, 1947], 245).

82. Many other features, in "Impressions de route en automobile" as well as elsewhere, bespeak the subtle ways in which Proust can be said to invert, even make fun of, Ruskin's antimodern stance. When, for example, the motorists are forced to stop in Lisieux because of a mechanical problem, the narrator takes a moment to study the façade of the cathedral because, as he points out to the reader, Ruskin once wrote about it. Night has fallen, however, and Lisieux is pitch-black. Agostinelli therefore turns on the headlights and directs them at the cathedral's exterior, endowing the porch with a halo of electric lighting. Of course, nothing could be farther from Ruskin's ideal lighting conditions.

83. Proust, "La bénédiction du sanglier," in *CSB*, 202. The pastiche was found after Proust's death, and Kolb has dated it to 1909.

84. I owe this piece of information to Malcolm Bowie's *Proust Among the Stars* (London: HarperCollins, 1998), 89. See also his fine reflections on Proust and Giotto (88–90).

85. Things are more complicated, however, for it is clear that Ruskin's own brilliant reading of Giotto's treatment of the angels paved the way for Proust's aerotechnical imagery. In fact, although Ruskin is never mentioned by name in the Padua episode, some of Proust's boldest reflections are indebted to Ruskin: "There is noticeable here," Ruskin thus writes in *Giotto and His Works in Padua*, "as in all works of this early time, a certain confidence in the way in which the angels trust to their wings, very characteristic of a period of bold and simple conception. Modern science has taught us that a wing cannot be anatomically joined to a shoulder; and in proportion as painters approach more and more to the scientific, as distinguished from the contemplative state of mind, they put the wings of their angels on more timidly, and dwell with greater emphasis upon the human form, and with less upon the wings, until these last become a species of decorative appendage,—a mere *sign* of an angel. But in Giotto's time an angel was a complete creature, as much believed in as a bird; and the way in which it would or might cast itself into the air, and lean hither and thither upon its plumes, was as naturally apprehended as the manner of flight of a chough or a starling" (*WJR* 24:72; emphasis in the original).

86. In a similar vein, Bowie argues that a materialist view of art exists side by side with an idealist one in *Remembrance;* he explores in particular how Proust's materialism makes itself felt in the recurrent celebrations of the labor that goes into any work of art; see Bowie, *Proust Among the Stars*, 68–125.

87. Observing that the Martinville episode builds on the reversal of mobility and immobility, Catherine Millot suggests that such a reciprocal exchange of attributes

"may be the major figure of the *Remembrance*" ("The Real Presence," *October* 58 [fall 1991]: 132).

88. Jeffrey T. Schnapp has similarly observed that the reversal figure—the substitution of immobility for mobility—organizes a great deal of representations of speed and motion in the modernist period. It recurs, he writes, "in nearly every early description of accelerated motion and underlies the design of thrill rides in nineteenth-century amusement parks as well as their cinematic shows: namely, *a reversal of perspective such that it now appears as if it is the landscape that is in motion and not the traveler; or, rather, that the landscape is in motion for the traveler*" ("Crash [Speed as Engine of Individuation]," *Modernism/modernity* 6, no. 1 [1999], 22; emphasis in the original). Schnapp argues that the figure can be traced as far back as eighteenth-century descriptions of mail-coach travel. My approach differs from his in the sense that I am concerned with the ways in which vehicles in motion—trains as well as automobiles—may serve as visual framing devices and how they, therefore, share an affinity with cinema. Also, I focus on early cinema and turn-of-the-century fairground attractions as *simulacra* of speed, that is, as *means of representation*. Combined, these two aspects constitute what I take to be a specifically late nineteenth- and early twentieth-century phenomenon.

89. As early as 1914, the French writer and painter Jacques-Emile Blanche, in a review of *Swann*, remarked that "Proust has not so much been keeping a journal as amusing himself with a kind of cinema film, reconstituting the sequences and *posing* in it himself for several characters, throwing, as the whim takes him, the cloak of one around the shoulders of another, or even wearing it himself" (*Marcel Proust: The Critical Heritage*, ed. Leighton Hodson [London: Routledge, 1989], 116). Another critic, reviewing *La Prisonnière* for *The Times* in 1924, also underscores the cinematic parallel: "No other author that I have ever read comes near Proust in the photographic, or rather kinematographic delineation of a big social 'crush,' with its tittle-tattle and back-biting, the snobbery and insolent condescension of its 'great ladies,' its occasional epigrams, its bowing and scraping *super ignes suppositos*, its stifling and noxious atmosphere, its glimpses of queer episodes behind *portières* and in remote corridors" (ibid., 269). Benjamin Crémieux, finally, observed that "Proust has been described as depicting life in *slow motion* and this formula has had a certain success. For all that, it is wrong; it is other novelists who depict life *speeded up*" (ibid., 289–90).

90. Among the few attempts at establishing parallels between Proust's style and cinematic modes of representation, in particular so-called travelings and close-ups, are Jacques Bourgeois, "Le cinéma à la recherche du temps perdu," *Revue du cinéma* 1, no. 3 (December 1946): 18–38. Bourgeois even argues that Proust anticipates a more "mature" form of narrative cinema—mature because psychologically inflected. See also Jacques Nantet's brief inventory of cinematographically influenced episodes in *A la recherche*, "Marcel Proust et la vision cinématographique," *Revue des lettres modernes* 5, nos. 36–38 (1958): 307–12. According to Nantet, the Martinville episode and the narrator's car ride to La Raspelière demonstrate how the "lens moves with the narrative." Renate Hörisch-Helligrath has similarly proposed that the description of the car ride is analogous to techniques of cinematic production, although in her essay, too, the suggestion remains undeveloped; see Hörisch-Helligrath, "Das deutende Auge:

Technischer Fortschritt und Wahrnehmungsweise in der *Recherche*," in *Marcel Proust: Motiv und Verfahren*, 22.

91. Jacques Rivière was arguably the first critic to draw attention to the affinity between Proust's style and cubist techniques; see *Marcel Proust et Jacques Rivière: Correspondance, 1914–1922*, ed. Philip Kolb (Paris: Plon, 1955), 264; and Rivière's *Quelques progrès dans l'étude du cœur humain (Freud et Proust)*, Les cahiers d'occident, no. 4 (Paris: Librairie de France, 1926), 83. See also Claude Gandelman, "Proust as a Cubist," *Art History* 2 (September 1979): 355–63; Hohl, "Marcel Proust in neuer Sicht"; P. A. Jannini, *Proust e le avanguardie*, in *Proustiana: Atti del Convegno internazionale di studi sull'opera di Marcel Proust* (Padua: Liviana, 1971), 113–22; Paola Placella Sommella, *Marcel Proust e i movimenti pittorici d'avanguardia* (Rome: Bulzoni, 1982); and Johnson, "Proust's 'Impressionism,' " 48–49.

92. Carter's notion; see Carter, *Proustian Quest*, 68–69, 73–78. See also Hohl, "Marcel Proust in neuer Sicht", 68–69; Bal, *Mottled Screen*, 227–34; Brassaï, *Marcel Proust sous l'emprise de la photographie*, 145–49.

93. There is more to the episode: the question of Saint-Loup's homosexuality. See Bal's reading of the episode in her *Mottled Screen*, 230–34.

94. F. T. Marinetti, "Futurist Painting: Technical Manifesto," in *Theories of Modern Art: A Source Book by Artists and Critics*, ed. Herschel B. Chipp (Berkeley: University of California Press, 1968), 289.

95. Quoted in Ruskin's *Modern Painters;* see *WJR* 5:369.

4. The Aesthetics of Immediacy

1. In preparing this chapter, I have consulted the following bibliographies and works of reference: Robert H. Deming, *A Bibliography of James Joyce Studies*, 2d ed., rev. and enl. (Boston: G. K. Hall, 1977); Thomas F. Staley, *An Annotated Critical Bibliography of James Joyce* (New York: St. Martin's Press, 1989); Thomas Jackson Rice, *James Joyce: A Guide to Research* (New York: Garland, 1982); Derek Attridge, ed., *The Cambridge Companion to James Joyce* (Cambridge: Cambridge University Press, 1990); Robert H. Deming, ed., *James Joyce: The Critical Heritage*, 2 vols. (London: Routledge, 1970).

2. Man Ray had decided to substitute the airbrush for the traditional brush, thus adapting a technique normally used in commercial artwork. As the reception of his new work testifies, Man Ray's choice of technique was a symbolically charged act. *Admiration of the Orchestrelle for the Cinematograph,* like his other airbrush paintings, met with harsh criticism when first exhibited, precisely because the technique was associated with commercial work. In other words, Man Ray's work invited a critical apparatus framed by the opposition between the mechanical and the human, the commercial and the spiritual. Man Ray, therefore, was perceived as having "vulgarized and debased art." See Man Ray, *Self Portrait* (Boston: Little, Brown, 1963), 76. See also Roland Penrose, *Man Ray* (Boston: New York Graphic Society, 1975), 45–48.

3. Maxim Gorky, review of the Lumière program, in *Kino: A History of the Russian and Soviet Film,* ed. Jay Leyda, 3d ed. (Princeton, N.J.: Princeton University Press, 1983), 407.

4. Frank Budgen, *James Joyce and the Making of Ulysses* (Bloomington: Indiana University Press, 1960), 21. In a letter to Carlo Linati, Joyce explained the idea of the

novel in similar terms, calling *Ulysses* "the cycle of the human body" (*Letters of James Joyce*, ed. Stuart Gilbert [New York: Viking, 1966], 1:146).

5. Joyce, letter to Carlo Linati. The passage reads: "[*Ulysses*] is an epic of two races (Israelite—Irish) and at the same time the cycle of the human body as well as a little story of a day (life). The character of Ulysses always fascinated me—even when a boy. Imagine, fifteen years ago I started writing it as a short story for *Dubliners!* For seven years I have been working at this book—blast it! It is also a sort of encyclopaedia. My intention is to transpose the myth *sub specie temporis nostri*. Each adventure (that is, every hour, every organ, every art being interconnected and interrelated in the structural scheme of the whole) should not only condition but even create its own technique. Each adventure is so to say one person although it is composed of persons—as Aquinas relates of the angelic hosts" (1:146–47).

6. In December 1921, Joyce's friend Valéry Larbaud had given a lecture on Joyce's coming novel, emphasizing among other things the proposed correspondences between *Ulysses* and *The Odyssey*. When the novel came out, Joyce, who had supported Larbaud's initiative, complained that no critic had picked up the thread. As Richard Ellmann explains, Joyce encouraged another friend, T. S. Eliot, to write an article that pursued the parallel. Eliot then published a review in *The Dial* where he maintained that Joyce's idea that antiquity should reflect contemporaneity had the "importance of a scientific discovery" (Ellmann, *James Joyce*, rev. ed. [Oxford: Oxford University Press, 1983], 527). Larbaud's and Eliot's articles are but two examples, yet they show just how successfully Joyce aligned the interpretive horizons of his critics with his own. The Homeric parallel quickly became an interpretive *doxa* in much of Joyce scholarship, and in this respect, Stuart Gilbert's 1931 study of *Ulysses* was instrumental. Gilbert got Joyce's permission to reproduce the famous chart in his classic study, which fleshes out the Greek parallels episode by episode. See Gilbert, *James Joyce's Ulysses* (New York: Vintage, 1955).

7. Budgen, *James Joyce and the Making of Ulysses*, 21.

8. References cite episode number, followed by line number. The reissued 1922 edition of *Ulysses* (Oxford: Oxford University Press, 1993) has been consulted for possible variants.

9. Budgen, *James Joyce*, 21.

10. Donald Theall has recently published two timely and comprehensive studies that situate Joyce's *Ulysses* and *Finnegans Wake* in a historical context determined by technological change, particularly with regard to communications technology. At the same time, Theall explores how current theories of communication are prefigured in *Ulysses*. See Theall, *Beyond the Word: Reconstructing Sense in the Joyce Era of Technology, Culture, and Communication* (Toronto: University of Toronto Press, 1995), as well as his *James Joyce's Techno-Poetics* (Toronto: University of Toronto Press, 1997). Apart from Theall's studies, the issue of technology has received little attention in literary scholarship devoted to *Ulysses*. Hugh Kenner's writings on Joyce offer an exception, usefully tracing how the advent of modern technology is reflected on the level of motifs; see Kenner, *The Mechanic Muse* (New York: Oxford University Press, 1987), as well as his "Notes Towards an Anatomy of 'Modernism,' " in *A Starchamber Quiry: A James Joyce Centennial Volume, 1882–1982*, ed. Edmund L. Epstein (London: Methuen, 1982), 3–42. Derrida, in a commentary on *Ulysses*, offers a series of suggestive remarks con-

cerning telephony and gramophony, less as historical phenomena, however, and more as emblems of the spacing of meaning which, on Derrida's view, is what enables all production of meaning, even before the advent of telephony and gramophony. See Derrida, "Ulysses Gramophone: Hear Say Yes in Joyce," trans. Tira Kendall, in *Acts of Literature*, ed. Derek Attridge (New York: Routledge, 1992), 253–309. As for Joyce and cinema, finally, no book-length study exists, but two studies of film and literary modernism contain a wealth of perceptive commentaries on Joyce's *Ulysses;* see Alan Spiegel, *Fiction and the Camera Eye: Visual Consciousness in Film and the Modern Novel* (Charlottesville: University Press of Virginia, 1976), esp. 71–82, 91–97, 109–15, 136–50, 164–74; and Keith Cohen, *Film and Fiction: The Dynamics of Exchange* (New Haven, Conn.: Yale University Press, 1979), esp. 112–13, 126, 146–56, 172–92, 197–206. See also the essays listed in note 23.

11. Except for a few articles on the "Proteus" episode (see note 46 below), the theme of perception in *Ulysses* has received little attention in Joyce criticism. Sheldon Brivic has devoted a book-length study to "the Lacanian gaze" in Joyce, but does not discuss visual perception per se, much less explore the obsession with eyes and visuality in *Ulysses;* see Brivic, *The Veil of Signs: Joyce, Lacan, and Perception* (Urbana: University of Illinois Press, 1991). Roy Gottfried has discussed how Joyce, by having changed the place and order of letters in words, "trains the reader's eye"; this discussion, however, also neglects the issue of perception as such; see Gottfried, "Reading Figather: Tricks of the Eye in 'Ulysses,' " *James Joyce Quarterly* 25, no. 4 (summer 1988): 465–74.

12. Franco Moretti, in a suggestive analysis of Bloom's deciphering of urban sensory stimuli, proposes that Joyce's stream-of-consciousness technique is an eminently *metropolitan* technique. Using Georg Simmel's classic essay on the modern city as a point of departure, Moretti argues that Joyce's stream-of-consciousness technique offers a stylistic solution to a historical problem: how to match and reproduce life in the metropolis, notably its visual worlds. Moretti thus suggests that the interior monologue accommodates what he calls the "stimuli of modernity," "keeping them all present, in the foreground, without losing a single one of them" (Moretti, "*Ulysses* and the Twentieth Century," in *Modern Epic: The World-System from Goethe to García Márquez*, trans. Quintin Hoare [New York: Verso, 1996], 123–229; quotation, 162). What Moretti fails to discuss, however, is that metropolitan perception is an overdetermined phenomenon. It is to be understood as an umbrella term, gathering up a diverse array of stimuli that need to be broken down into specific components: stimuli produced by, for example, technologies of velocity such as the tram; mnemonic devices such as newspapers and advertisements; and technologies of reproduction such as photography, cinematography, telephony, and phonography. Also, it is important to note that the urban space that emerges in *Ulysses* is not an anonymous city, nor is it peopled by crowds; and this sets Joyce's Dublin apart from, for example, the cities in Baudelaire, T. S. Eliot, Virginia Woolf, and Robert Musil.

13. Viktor Shklovsky, *Zoo, or Letters Not about Love* [1923], trans. Richard Sheldon (Ithaca, N.Y.: Cornell University Press, 1971), 80–82. "In the primitive novel," Shklovsky writes, "the hero is a vehicle for connecting the parts. When works of art are undergoing change, interest shifts to the connective tissue" (80).

14. Joseph Frank, "Spatial Form in Modern Literature," in *The Idea of Spatial Form* (New Brunswick, N.J.: Rutgers University Press, 1991), 21.

15. Spiegel, *Fiction and the Camera Eye*, 64.

16. On Proust's theory of sensory reification, see the Introduction to the present work.

17. Virginia Woolf, *The Years* (London: Penguin, 1968), 263. The last section in particular, "Present Day," abounds with representations of acoustic experiences, including autonomous voices.

18. *Ulysses* offers many examples of autonomous or reified voices: "Have you the key? a voice asked" (*U* 1.322); "She bows her old head to a voice that speaks to her loudly [. . .]. And to the loud voice that now bids her be silent with wondering unsteady eyes" (*U* 1.418–23); "A voice, sweettoned and sustained, called to him from the sea" (*U* 1.741); "Bantam Lyons raised his eyes suddenly and leered weakly./—What's that? his sharp voice said" (*U* 5.536); "With awe Mr Power's blank voice spoke" (*U* 6.921–22); "The inner door was opened violently and a scarlet beaked face, crested by a comb of feathery hair, thrust itself in. The bold blue eyes stared about them and the harsh voice asked:—What is it?" (*U* 7.344–47); "The noise of two shrill voices, a mouthorgan, echoed in the bare hallway from the newsboys squatted on the doorsteps:/—*We are the boys of Wexford/Who fought with heart and hand*" (*U* 7.425–28); "In a giggling peal young goldbronze voices blended, Douce with Kennedy your other eye" (*U* 11.158–59); "Miss voice of Kennedy answered, a second teacup poised, her gaze upon a page:—No. He was not. Miss gaze of Kennedy, heard, not seen, read on" (*U* 11.237–40); "The voice of the mournful chanter called to dolorous prayer" (*U* 11.1132–33); "By screens of lighted windows, by equal gardens a shrill voice went crying, wailing: *Evening Telegraph, stop press edition! Result of the Gold Cup races!*" (*U* 13.1173–75); "Then he looked up and saw the eyes that said or didn't say the words the voice he heard said, if you work.—Count me out, he managed to remark, meaning work. The eyes were surprised at this observation because as he, the person who owned them pro tem. observed or rather his voice speaking did, all must work, have to, together" (*U* 16.1146–51).

19. For a poststructuralist reading of the "rhetorical separation of voice" in *Ulysses*, see Scott W. Klein, *The Fictions of James Joyce and Wyndham Lewis: Monsters of Nature and Design* (Cambridge: Cambridge University Press, 1994), 74–79. Klein suggests that "voice and speech are no longer unproblematic vessels of 'presence' in these chapters, neither naturally accessible to the self nor unequivocal in their ability to express truth" (77).

20. On the differentiation of the eye and the ear in *Finnegans Wake*, and its affinities with cinema, see Klein, *The Fictions of James Joyce and Wyndham Lewis*, 174–77.

21. J. Mitchell Morse, for example, states that "in general Joyce favored time and the ear" (Morse, "Joyce and the Blind Stripling," *Modern Language Notes* 71, no. 7 [November 1956]: 498). Joseph E. Duncan suggests that "Joyce apparently devoted particular attention to the use of sound in *Ulysses* because of his own highly developed auditory sensitivity" (Duncan, "The Modality of the Audible in Joyce's *Ulysses*," *PMLA* 72, no. 1 [March 1957]: 290–91). On the relation of music and Joyce's writings, see Zack Bowen, *Musical Allusions in the Works of James Joyce* (Albany: State University of New York Press, 1974), as well as his *Bloom's Old Sweet Song: Essays on Joyce and Music* (Gainesville: University Press of Florida, 1995). See also the essays collected in *Picking Up Airs: Hearing the Music in Joyce's Text*, ed. Ruth H. Bauerle (Urbana: University of Illinois Press, 1993).

22. According to his friend Georges Borach, Joyce explained his work on the "Sirens" episode in the following terms. "I finished the Sirens chapter during the last few days. A big job. I wrote this chapter with the technical resources of music. It is a

fugue with all musical notations: *piano, forte, rallentando,* and so on. A quintet occurs in it, too, as in the *Meistersinger,* my favorite Wagner opera" (*Portraits of the Artist in Exile: Recollections of James Joyce by Europeans,* ed. Willard Potts [Seattle: University of Washington Press, 1979], 72).

23. Visuality in *Ulysses* has been theorized in two major ways: in terms of cubism and in terms of cinema. On the kinship between Joyce's means of representation and cubist techniques, see Heinz Brüggemann, "Bewegtes Sehen und literarisches Verfahren: James Joyces 'Ulysses' und der Kubismus," *Neue Rundschau* 102, no. 3 (1991): 146–59; Archie K. Loss, *Joyce's Visible Art: The Work of Joyce and the Visual Arts, 1904–1922* (Ann Arbor, Mich.: UMI Research Press, 1984); and Jo-Anna Isaak, "James Joyce and the Cubist Esthetic," *Mosaic* 14, no. 1 (winter 1981): 61–90. As for cinematic perspectives on Joyce, Sergei Eisenstein's writings have been instrumental. A fervent admirer of Joyce's *Ulysses,* Eisenstein focused on Joyce's montage-like techniques and his use of the interior monologue, arguing, for example, that *Ulysses,* like cinema, is characterized by a " 'physiologism' of detail." Since both cinema and Joyce's novel successfully explore closeup techniques, he proposed, both make the visual universe susceptible to detailed analysis, hence also producing essentially new visual worlds. See Eisenstein, "Literature and Cinema: Reply to a Questionnaire," in *Selected Works,* ed. and trans. Richard Taylor (London: British Film Institute, 1988), 1:95–99; "Laocoön," in *Selected Works,* ed. Michael Glenny and Richard Taylor, trans. Michael Glenny (London: British Film Industry, 1991), 2:193–202; and "Achievement," in *Film Form: Essays in Film Theory,* ed. and trans. Jay Leyda (San Diego: Harcourt Brace Jovanovich, 1977), 179–94. For a transcript of the lecture in which Eisenstein introduces *Ulysses* to students at the Institute of Cinematography in Moscow, see Emily Tall, "Eisenstein on Joyce: Sergei Eisenstein's Lecture on James Joyce at the State Institute of Cinematography, November 1, 1934," *James Joyce Quarterly* 24, no. 2 (winter 1987): 133–42. For a discussion of Eisenstein's interest in Joyce's work, see Gösta Werner, "James Joyce and Sergej Eisenstein," trans. Erik Gunnemark, *James Joyce Quarterly* 27, no. 3 (spring 1990), 491–507. See also Spiegel, *Fiction and the Camera Eye,* esp. 71–81; Keith Cohen, *Film and Fiction: The Dynamics of Exchange* (New Haven, Conn.: Yale University Press, 1979), esp. 126–56; Mary Parr, *James Joyce: The Poetry of Conscience* (Milwaukee, Wis.: Inland Press, 1961); Austin Briggs, " 'Roll Away the Reel World, the Reel World': 'Circe' and Cinema," in *Coping with Joyce: Essays from the Copenhagen Symposium,* ed. Morris Beja and Shari Benstock (Columbus: Ohio State University Press, 1989), 145–56; Thomas L. Burkdall, "Cinema Fakes: Film and Joycean Fantasy," in *Joyce in the Hibernian Metropolis,* ed. Morris Beja and David Norris (Columbus: Ohio State University Press, 1996), 260–68; and Ruth Perlmutter, "Joyce and Cinema," *boundary 2* 6, no. 2 (winter 1978): 481–502.

24. Wyndham Lewis, *Tarr. The 1918 Version,* ed. Paul O'Keeffe (Santa Rosa, Calif.: Black Sparrow Press, 1990), 98. For an analysis of Lewis's style, its brutally aestheticizing impulse and its dismantling of human gestures, see Fredric Jameson, *Fables of Aggression: Wyndham Lewis, the Modernist as Fascist* (Berkeley: University of California Press, 1981), 25–86. On how Lewis's machine aesthetic makes itself felt in *Tarr,* specifically on the level of narrative form, see Michael Wutz, "The Energetics of *Tarr:* The Vortex-Machine Kreisler," *Modern Fiction Studies* 38, no. 4 (winter 1992): 845–69.

25. See the Introduction to the present work.

26. James Joyce, *A Portrait of the Artist as a Young Man* (New York: Viking, 1967), 171.

27. On Joyce's interest in the detail, see Karen Lawrence, *The Odyssey of Style in "Ulysses"* (Princeton, N.J.: Princeton University Press, 1981), 186–89; quotation, 187.

28. Joseph Conrad, preface to *The Nigger of the "Narcissus"* (London: Penguin, 1988), xlix.

29. Shklovsky, "Art as Device," in *Theory of Prose*, trans. Benjamin Sher (Elmwood Park, Ill.: Dalkey Archive Press, 1990), 6.

30. Joyce was keenly interested in cinema and, unlike Mann and Proust, appears not to have considered film as antithetical to what he thought of as art. In fact, Joyce established the first permanent picture theater in Ireland. The Volta opened in 1909, featuring "The First Paris Orphanage," "La Pourponnière," and "The Tragic Story of Beatrice Cenci," including a "little string orchestra" playing "charmingly," as a critic noted. The enterprise, however, proved to be less lucrative than Joyce had hoped. For a history of Joyce's short cinema venture, see Gösta Werner, "James Joyce, Manager of the First Cinema in Ireland," *Moderna Språk* [Modern Language Teachers' Association of Sweden] 76, no. 2 (1982): 131–42. See also Ellmann, *James Joyce*, 303–4, 310–11.

31. Gilles Deleuze, *Cinema 1: The Movement-Image*, trans. Hugh Tomlinson and Barbara Habberjam (Minneapolis: University of Minnesota Press, 1991), 14–15.

32. Briggs, " 'Roll Away the Reel World, the Reel World,' " 146.

33. On the railway as a vehicle of perception, see Wolfgang Schivelbusch, *The Railway Journey: The Industrialization of Time and Space in the Nineteenth Century* (Berkeley: University of California Press, 1986). On the railway as a proto-cinematic phenomenon, see Lynne Kirby, *Parallel Tracks: The Railroad and Silent Cinema* (Durham, N.C.: Duke University Press, 1997).

34. László Moholy-Nagy, *Vision in Motion* (Chicago: Paul Theobald, 1947), 343; emphasis in the original.

35. Walter Benjamin, "The Work of Art in the Age of Mechanical Reproduction," in *Illuminations*, trans. Harry Zohn (New York: Schocken, 1988), 237.

36. Italo Calvino, "Gustave Flaubert, *Trois Contes*," in *Why Read the Classics?* trans. Martin McLaughlin (New York: Pantheon Books, 1999), 151–52. Although Calvino does not state it explicitly, the passage intimates that visibility in the novel coincides with the advent of photography. On the relations of photography and the nineteenth-century British novel, see Nancy Armstrong, *Fiction in the Age of Photography: The Legacy of British Realism* (Cambridge, Mass.: Harvard University Press, 1999).

37. See Werner, "James Joyce and Sergej Eisenstein," 494. For Eisenstein's writings on Joyce, see note 23 above.

38. Fernand Léger, "The Machine Aesthetic: Geometric Order and Truth," in *Functions of Painting*, ed. Edward F. Fry, trans. Alexandra Anderson (New York: Viking, 1973), 65.

39. Léger, "The Machine Aesthetic: The Manufactured Object, the Artisan, and the Artist," in *Functions of Painting*, 52–53.

40. Léger, "Critical Essay on the Plastic Quality of Abel Gance's *The Wheel*," in *Functions of Painting*, 22; emphasis in the original.

41. Both Léger and Vertov, incidentally, present closeups of the human eye, blinking and watching, now a specular world unto itself to be studied and looked at, not unlike Dr Behrens's examination of Hans Castorp's means of vision. In Léger and Vertov, however, the eye turns into an object for aesthetic contemplation, and is ultimately to

be understood as a meta-aesthetic commentary—a commentary on the prominence of the visual methods used.

42. Dziga Vertov, "The Man with a Movie Camera," in *Kino-Eye: The Writings of Dziga Vertov*, ed. Annette Michelson, trans. Kevin O'Brien (Berkeley: University of California Press, 1984), 283; emphasis in the original.

43. Vertov, "From Kino-Eye to Radio-Eye," in *Kino-Eye*, 87.

44. Ibid., 87–88.

45. Shklovsky, "Art as Device," 6; emphasis in the original. The translator offers a neologism, "enstrangement," but I have kept to the widely used "estrangement" in order to avoid confusion. On Shklovsky's notion of *ostraniene*, see Victor Erlich, *Russian Formalism: History—Doctrine* (The Hague: Mouton, 1955), 57, 149–54.

46. On the philosophical components in "Proteus," see Duncan, "The Modality of the Audible in Joyce's *Ulysses*," 286–95; John Killham, "'Ineluctable Modality' in Joyce's *Ulysses*," *University of Toronto Quarterly* 34, no. 3 (1965): 269–89; and Pierre Vitoux, "Aristotle, Berkeley, and Newman in 'Proteus' and *Finnegans Wake*," *James Joyce Quarterly* 18, no. 2 (1981): 161–75.

47. Cohen, *Film and Fiction*, 152.

48. André Bazin, "Painting and Cinema," in *What Is Cinema?* ed. and trans. Hugh Gray, 2 vols. (Berkeley: University of California Press, 1967), 1:166. See also "Theater and Cinema—Part Two," in *What Is Cinema?* 1:105.

49. Susan Stewart has commented upon Bloom's voyeurism and the photographic/pornographic nature of the representation of what he sees; see Stewart, *On Longing: Narratives of the Miniature, the Gigantic, the Souvenir, the Collection* (Durham, N.C.: Duke University Press, 1993), 115–16.

50. Other examples of how Gertrude's gaze carries a hermeneutics that is projected onto Bloom's exterior include: "She could almost see the swift answering flash of admiration in his eyes that set her tingling in every nerve" (*U* 13.513–14); "Passionate nature though he was Gerty could see that he had enormous control over himself. One moment he had been there, fascinated by a loveliness that made him gaze, and the next moment it was the quiet gravefaced gentleman, selfcontrol expressed in every line of his distinguishedlooking figure" (*U* 13.539–43); "If ever there was undisguised admiration in a man's passionate gaze it was there plain to be seen on that man's face. It is for you, Gertrude MacDowell, and you know it" (*U* 13.564–67); "Whitehot passion was in [his] face, passion silent as the grave, and it had made her his" (*U* 13.691–92).

51. Gaston Picard, interview with Apollinaire [1917], in Apollinaire, *Œuvres en prose complètes*, ed. Pierre Caizergues and Michel Décaudin, 3 vols. (Paris: Gallimard, 1991), 2:989.

52. Joyce's interest in Wagner is well known and well documented, and Wagnerian references in *Ulysses* abound. For example, the artist-hero Stephen Dedalus is partly modeled on Wagner's Siegfried, as Joyce himself confirmed. And like Thomas Mann, Joyce attempted to translate the musical leitmotiv into a novelistic device, the kidney being one example, the lemony soap another. On Joyce and Wagner's music, see William Blissett, "James Joyce in the Smithy of His Soul," in *James Joyce Today: Essays on the Major Works*, ed. Thomas F. Staley (Bloomington: Indiana University Press, 1966), 96–134. Timothy Martin has devoted a book-length study to the influence of Wagner upon Joyce's works, arguing that although Joyce's persistent interest in Wag-

ner's music dramas accounts for an important source of influence, the most crucial shaping force as far as Joyce's writings are concerned is literary Wagnerism. Martin discusses writers whose works were essential to Joyce, such as Mallarmé, who explicitly set out to outdo Wagner in the realm of poetry and projected the idea of the all-encompassing book; Edouard Dujardin, who edited the *Revue wagnérienne* and inspired Joyce's interior monologue; and Arthur Symons, who wrote *Symbolist Movement in Literature,* in which Wagner assumes an important role. See Martin, *Joyce and Wagner: A Study of Influence* (Cambridge: Cambridge University Press, 1991). Interestingly, both Blissett and Martin suggest that Joyce's interior monologue has a Wagnerian provenance; Joyce explained that he was inspired by Dujardin's *Les lauriers sont coupés,* and Dujardin, in his turn, explained that he was influenced by Wagner's continuous melody. See Blissett, "James Joyce in the Smithy of His Soul," 113; and Martin, "Joyce and Literary Wagnerism," in *Picking Up Airs,* 120.

53. For a reading of *Ulysses* as a "newspaper landscape," see Marshall McLuhan, "Joyce, Mallarmé, and the Press," *Sewanee Review* 62, no. 1 (winter 1954): 38–55.

54. For an in-depth study of memory in Joyce, see John S. Rickard, *Joyce's Book of Memory: The Mnemotechnic of "Ulysses"* (Durham, N.C.: Duke University Press, 1999). Rickard, however, does not consider the theme of memory in the context of mnemotechnic devices such as the camera and the gramophone.

55. Avital Ronell, *The Telephone Book: Technology, Schizophrenia, Electric Speech* (Lincoln: University of Nebraska Press, 1989), 438, n. 110. Ronell cites Proust (*Cities of the Plain*), Mark Twain (*A Connecticut Yankee in King Arthur's Court*), and Max Brod (the poem "Telephon," in which the telephone booth emerges as a grave), as well as a 1886 entry for a U.S. patent discussing the so-called phantom circuit.

56. Walter Rathenau, "Die Resurrection Co." [1898], in *Gesammelte Schriften* (Berlin: S. Fischer, 1925), 4:339–49. The thrust of the story is, among other things, a subtle critique of all things mechanized (as well as things American and capitalist). The undertaker, Mr Elihu Hannibal I. T. Gravemaker, has successfully applied the conveyor-belt principle to the funeral ceremony. An electric tram speeds the coffin out to the cemetery; a crane lifts the coffin off the tracks and into the machine-dug grave; recorded words of consolation may be had for twenty cents; and so on. When it turns out that a "customer" has been buried alive, Mr Gravemaker decides to prevent further scandals. He accepts Dacota- and Central-Resurrection Telephone and Bell Co.'s suggestion that all graves be endowed with wires, telephones, and bells, but runs into new problems: an entire network of noisy underground communication spreads out in the most solemn of places. I am grateful to Aris Fioretos for bringing Rathenau's short story to my attention. See Fioretos's discussion of Rathenau in *En bok om fantomer* (Stockholm: Norstedts, 1996), 74–77.

57. See also the "Eumaeus" episode, where Bloom reflects upon photography in aesthetic terms. While peering at a photograph of Molly, he offers a critical judgment on whether photography may do justice to his wife's physical appearance. Comparing photography to Greek sculpture, he concludes that the latter is far superior in its representational solutions. "Whereas no photo could because it simply wasn't art in a word" (*U* 16.1454–55).

58. On reification and fragmentation in Joyce, see Fredric Jameson, " 'Ulysses' in

History," in *James Joyce: Modern Critical Views,* ed. Harold Bloom (New York: Chelsea House, 1986), 173–88.

59. See, for example, Gilbert, *James Joyce's Ulysses.*

60. Colin MacCabe, *James Joyce and the Revolution of the Word* (New York: Barnes and Noble, 1979), 104.

61. Stewart, *On Longing,* 22.

62. T. S. Eliot, "*Ulysses,* Order and Myth," *Dial* 75 (November 1923): 480–83. Reprinted in *James Joyce: Two Decades of Criticism,* 198–202. "If [*Ulysses*] is not a novel," Eliot writes, "that is simply because the novel is a form which will no longer serve; it is because the novel, instead of being a form, was simply the expression of an age which had not sufficiently lost all form to feel the need of something stricter" (201). For a discussion of how *Ulysses* "denies the validity of genres," see A. Walton Litz, "The Genre of 'Ulysses,' " in *The Theory of the Novel,* ed. John Halperin (New York: Oxford University Press, 1974), 109–20.

63. Conrad accordingly asserted that fiction, like all art, "must strenuously aspire to the plasticity of sculpture, to the colour of painting, and to the magic suggestiveness of music—which is the art of arts" (Conrad, preface to *The Nigger of the "Narcissus,"* xlix).

64. Moholy-Nagy, *Painting, Photography, Film,* trans. Janet Seligman (London: Lund Humphries, 1969), 16–19.

65. Marcel Proust, letter to Mme Straus, 6 November 1908, *Selected Letters,* ed. Philip Kolb, trans. Terence Kilmartin (New York: Oxford University Press, 1989), 2:408–9.

66. Theodor W. Adorno, *Aesthetic Theory,* trans. Robert Hullot-Kentor (Minneapolis: University of Minnesota Press, 1997), 204.

67. Ellmann, *James Joyce,* 397.

Coda

1. Herman Hesse, *Steppenwolf,* trans. Basil Creighton, rev. Joseph Mileck and Horst Frenz (New York: Henry Holt, 1990), 180.

2. Theodor W. Adorno, "Short Commentaries on Proust," in *Notes to Literature,* trans. Shierry Weber Nicholsen, 2 vols. (New York: Columbia University Press, 1991), 1:178.

3. Walter Benjamin, "The Image of Proust," in *Illuminations,* trans. Harry Zohn (New York: Schocken, 1988), 205.

4. Virginia Woolf, *A Moment's Liberty: The Shorter Diary,* ed. Anne Olivier Bell (London: Hogarth Press, 1990), 149. In her diary, Woolf also recorded that Eliot had remarked that *Ulysses* "left Joyce himself with nothing to write another book on." Eliot thought, she added, "that Joyce did completely what he meant to do" (149).

5. On how Joyce explores the language of advertising, see Jennifer Wicke, "Scene of Writing in *Ulysses,*" in *Advertising Fictions: Literature, Advertisement, and Social Reading* (New York: Columbia University Press, 1988), 120–69. See also Alfred Paul Berger, "James Joyce, Adman," *James Joyce Quarterly* 3 (1965): 25–33.

6. Hugh Kenner has stressed that Dublin at the time had the most advanced electric tram system in Europe; see Kenner, "Notes Toward an Anatomy of 'Modernism,' "

in *A Starchamber Quiry: A James Joyce Centennial Volume, 1882–1982*, ed. E. L. Epstein (New York: Methuen, 1982), 5.

7. Kenner, *The Mechanic Muse* (New York: Oxford University Press, 1987), 11.

8. In reaching this conclusion, I am influenced by Jonathan Crary's inquiry into the construction of the observer in the early nineteenth century. What Crary argues with regard to vision may usefully be extended to aesthetic experience at large, a historical development that then culminates in the early twentieth century: "Once vision became located in the empirical immanence of the observer's body, it belonged to time, to flux, to death. The guarantees of authority, identity, and universality supplied by the camera obscura are of another epoch" (*Techniques of the Observer: On Vision and Modernity in the Nineteenth Century* [Cambridge, Mass.: MIT Press, 1990], 24).

9. For Hegel's hierarchy of the senses as organs of artistic enjoyment, see *Hegel's "Aesthetics": Lectures on Fine Art*, trans. T. M. Knox, 2 vols. (Oxford: Oxford University Press, 1975), esp. 1:38–39, 2:621–23; quotation, 2:621.

10. On the eye and the ear as aesthetic organs, and the division of labor between them, see also Richard Wagner, "The Art-Work of the Future," in *The Art-Work of the Future*, trans. William Ashton Ellis (Lincoln: University of Nebraska Press, 1993), 91–94.

11. Jacques Derrida, "The Pit and the Pyramid: Introduction to Hegel's Semiology," in *Margins of Philosophy*, trans. Alan Bass (Chicago: University of Chicago Press, 1982), 71–108.

12. Roland Barthes, *Camera Lucida: Reflections on Photography*, trans. Richard Howard (London: Fontana, 1982), 21, 27.

Index

Numbers in italics refer to illustrations.

Camartin, Iso, 210n. 27
Camera Lucida (Barthes), 195–96
camera obscura, 108, 114
camera. *See* photography
canon, 26–27
capitalism: Benjamin on, 50, 52–53; effects of, on art and culture, 36–40; imperialist stage of, 25; and modernism, 27–28, 30
Carey, John, 199n. 19
Carnet de 1908 (Proust), 221n. 69
Carter, William C., 97, 124, 222n. 74, 225n. 92
Cartwright, Lisa, 212n. 50
causality, 3, 11, 39. *See also* constitution; determinism; mediation
Chanan, Michael, 213n. 59
Chartier, Roger, 38
Chevrier, Jean-François, 219n. 46
chronophotography, 2, 5, 22, 56; Bergson's critique of, 102–6; influence on art and aesthetics, 100–103; invention of, 97–100; and movement, 98–107; and Proust, 94, 143–45; and relativization of human vision, 98–100, 104–7; and science of labor, 97
cinematic vision: Bazin's theory of, 176; in *Remembrance*, 130–33, 138–41; in *Ulysses*, 164–67, 185, 193. *See also* cinematography
cinematography, 2, 5, 22–23, 39, 45–46; Bergson's critique of, 104–5; compared to literature, 120; Deleuze on, 165–66; Gorky on, 149; and illusion of movement, 127–28, 141; influence of chronophotography on, 97, 104; invention of, 104; Léger on, 167–71; in *Magic Mountain*, 84–85, 190; in Man Ray, 147–48; and memory, 119–20; and modernist novel, 167, 193; and painting, 147–48, 168; and perceptual division of labor, 148–49; in Proust, 94, 119–21, 123, 141; and transformation of vision, 167–71; Tolstoy on, 128. *See also* cinematic vision
cinéorama, 127–28
civilization: as opposed to *Kultur*, 70–71
Clark, T. J., 198n. 7
class: as signifying system, 66–71
Claude glass, 116–17, 136. *See also* optical devices
Clayton, Jay, 201n. 34
Cocking, John M., 217n. 32
Cohen, Keith, 133, 218n. 33, 226n. 10, 229n. 23
Compagnon, Antoine, 199n. 16
Conrad, Joseph, 4, 10, 20, 22, 164, 186, 233n. 63; *Lord Jim,* 20, 159

constitution, 2, 10. *See also* mediation
Contre Sainte-Beuve (Proust), 221n. 69
Cook, David A., 215n. 3
Cozea, Angela, 200n. 29
Crary, Jonathan, 55–56, 63, 201n. 38, 234n. 8
Crémieux, Benjamin, 224n. 89
crowds: Benjamin on, 52–53
cubism: and technology, 2
culture industry. *See* mass culture
culture: German concept of, 70–71, 86, 190
Curtius, Ernst Robert, 202n. 10; on Mann, 86; on Proust, 132–33; 200n. 53, 222n. 74

Dagognet, François, 216n. 10
De Man, Paul, 33, 206n. 47
Death in Venice (Mann), 27–28, 57
Debord, Guy, 11
decadence, 32
defamiliarization. *See* estrangement
Degas, Edgar, 98, 101–2, 216n. 16
Deleuze, Gilles: on the cinematographic image, 165–66; on the machine, 41; *Proust and Signs,* 123, 192, 221n. 72
Demachy, Robert, 124, *125*
Deming, Robert H., 225n. 1
Demolder, Eugène: *L'Espagne en auto,* 126–27
Den Tandt, Christophe, 219n. 49
Derrida, Jacques, 26, 204n. 25, 226n. 10; on Hegel's aesthetics, 195
determinism, technological, 9, 39–43; Heidegger's critique of, 39; Latour's critique of, 41–43; Williams's critique of, 39–40. *See also* causality; mediation
diffusion model: Latour's critique of, 41–42
diorama, 62
discourse network: Kittler's theory of, 44–46
disembodiment of voice, 12–17, 157–58
division of labor: Marx on, 52; and chronophotography, 100
Doane, Mary Ann, 216n. 10
Dos Passos, John, 4
double, the: theme of, in *Remembrance,* 12–17; in *Ulysses,* 180, 182–83
Drucker, Johanna, 203n. 17
Duchamp, Marcel, 98, 102, *103*, 143
Dujardin, Edouard, 231n. 52
Duncan, Joseph E., 228n. 21, 231n. 46

Eagleton, Terry: on aesthetics, 5
ear. *See* hearing
Edison, Thomas A., 45, 198n. 8; and kinetoscope, 96–97
Eisenstein, Elizabeth L., 206n. 57

Marey, Etienne-Jules, 19, 56, 96, *100*, *101*;
Bergson's critique of, 102–6; on deperceptualization of science, 105–7; development of chronophotography, 97–101; influence on art and aesthetics, 100–103;
and Proust, 94, 143–45
Marinetti, F. T., 128, 186
Martin, Paul, 201n. 30
Martin, Timothy, 231n. 52
Marvin, Carolyn, 197n. 2
Marx, Karl: on the human senses, 11; on machine labor, 52
Marx, Leo, 212n, 53
masculinity: male gaze in modernist novel,
18, 159–60; male gaze in *Ulysses*, 171. *See
also* femininity; gender
mass culture: Adorno and Horkheimer's
view of, 35, 39; and bourgeois modernity,
32; and kitsch, 31; in *Magic Mountain*,
84–85; and modernist art and literature,
7–8, 28, 35, 40, 179; and technology, 7–8,
31, 35
Masten, Jeffrey, 204n. 25
Matisse, Henri, 136–37, *137*, 223n. 80
Matter, Harry, 207n. 1
McFarlane, James, 25, 199n. 18
McLuhan, Marshall, 11, 16, 232n. 53
mediation: Benjamin's theory of, 48–54; between modernism and modernity, 44–46;
as translation of signifying systems,
42–43. *See also* causality; constitution;
determinism
medicine, 21; Foucault on the history of,
73–74; in Mann, 72–75, 78–84, 89–90; and
Marey, 97–100; and X-ray technology, 74,
78–79
Megay, Joyce N., 215n. 1
memory: affected by technology, 181–83;
Bergson's theory of, 51; compared to photography, 97, 118–20; in *Remembrance*, 51,
94, 96–97, 118–20; in *Ulysses*, 180–84. *See
also* experience; temporality
Mendelssohn, Moses, 37
Millot, Catherine, 223n. 87
Milton, John, 36
Mirbeau, Octave, *La 628–E8*, 126–27, 131
modernism: Adorno and Horkheimer on, 25,
39; and aesthetics of autonomy, 23–24,
26, 30, 33–35, 38–39, 197n. 6; allegorical
readings of, 46–54; Berman on, 44; and
chronophotography, 100–103; criticism
and historiography of, 2, 4–5, 7–8, 23–25,
28–34, 43–54, 193, 198n. 8; definition of,

3–4, 6–9, 28–29; essentialist view of, 29;
etymological meaning of, 28; and feminist theory, 11, 18; and gender, 176–77;
high modernism, 6–7, 35, 39–40; historical-materialist theories of, 39–40; humanistic view of, 29–33; and imperial
capitalism, 25, 27–28; interpretive horizon of, 25–28; Kern on, 43–44; and mass
culture, 7–8, 32, 35, 40; and modernization, 7–10, 27, 30–31, 44, 49–54, 189–90,
194–96; poetry, 30–31; postmodern interpretations of, 25–28, 198n. 7; as reaction
to bourgeois modernity, 32–33; as reaction to industrial culture, 33; and representation of speed, 124–27; roots of, in
renaissance culture, 33; self-understanding of, 31–33; and sensory experience,
18–23, 51–53, 189–196; and style, 29, 33,
171–77; symbolical readings of, 47–48;
and technology, 10–11, 17–18, 21–23, 27,
38–40, 43–44, 53–54; and temporal experience, 9, 25, 32–33; Williams on, 39–40;
Žižek on, 26–27. *See also* avant-garde
modernist novel: abstraction of sensory experience in, 3–4, 15–17, 22–24, 185–86; and
authentic experience, 3, 23; Calvino on,
58, 167; and crisis of visibility, 167; encyclopedic character of, 58, 186–88; as
Gesamtkunstwerk, 3–4, 178–79, 185–186;
and technologies of perception, 189–96;
and technologies of speed, 189–196. *See
also* modernism; *Magic Mountain*; *Remembrance of Things Past*; *Ulysses*
modernity, 25–26; advent of, in Japan, 9–10;
and aesthetics, 37–39, 145–46; in Baudelaire, 45–54; Berman's view of, 44; Calinescu's view of, 32; Friedrich's view of,
30–31; Kittler's view of, 44–46; and modernism, 9–11, 30–34, 38–39; in Proust,
121–22, 146
modernization: and emergence of modern
aesthetics, 35–40, 44, 46, 189–90, 194–96;
Ruskin's critique of, 116
Moholy-Nagy, László, 7, 59, 223n. 81; on
Ulysses, 166–67; on modernism, 186
Moi, Toril, 213n. 56
Monet, Claude, 55, 92, 107, 109
Monnin-Hornung, Juliette, 217n. 32
Moretti, Franco, 199n. 19, 202n. 8; on bil-
dungsroman, 59–60; on Joyce, 227n. 12
Moritz, Karl Philipp, 37
Morse, J. Mitchell, 228n. 21
Moss, Howard, 221n. 72

Staley, Thomas F., 225n. 1
Steinman, Lisa M., 198n. 8
Steppenwolf (Hesse), 189, 214n. 55
stereoscope, 62–63
Stewart, Susan, 185, 231n. 49
Stock, Janet C., 215n. 1
Stubbs, Katherine, 200n. 25
style: as affected by technologies of perception, 22–23; and definition of modernism, 29; historical changes of, 186–87; Proust on, 186–87; and speed in Proust, 124–27, 130–41; in *Ulysses*, 23, 153, 160–63, 173, 178
sublime, the. *See* technological sublime, the
surrealism: machine in, 87; and technology, 2
Swahn, Sigbrit, 215n. 1
Swales, Martin, 208n. 12
symbol: as interpretive trope, 46–48
symbolism, 34
Symons, Arthur, 231n. 52
synaesthesia, 3–4, 88–89, 156, 174, 185

Tabbi, Joseph, 219n. 49
tactility, 74, 156, 175, 185, 194. *See also* senses, the
Tadié, Jean-Yves, 130, 200n. 28, 215nn. 1, 3
Tall, Emily, 229n. 23
Tanizaki, Junichiro, 9–10
Tarr (Lewis), 18, 159, 164
taste, 152, 175, 194. *See also* senses, the
Tate, Allen, 44
Taylor, Elizabeth Russell, 215n. 1
techne, 6, 24; as opposed to art, 35–36
technogram: Latour's theory of, 42–43; in *Magic Mountain*, 75, 78, 87–88
technological sublime, the: in *Magic Mountain*, 79–80; in *Remembrance*, 122, 212n. 53
technological uncanny, 182–83
technologies of perception, 3, 5; and aesthetics, 5–6; and cinematography, 46, 147–49; internalization of, in modernist novel, 21–23, 191–96; and knowledge of body, 19, 56, 63; in *Magic Mountain*, 55–57, 59; and objectivity, 19; as opposed to production technologies, 5; optical technology, 63–64; in *Remembrance*, 12–17; in Woolf, 17–18, 19. *See also* framing techniques; *Magic Mountain*; optics; *Remembrance of Things Past*; technologies of speed; *Ulysses*; *and entries for individual technologies of perception*

technologies of speed, 5, 17; and cinematography, 127–29; in modernist art and literature, 124–27, 190–91; and Proust, 94, 123–24, 130–41, 190, 192; in *Ulysses*, 191. *See also* automobile; framing techniques; railway; *Remembrance of Things Past*; speed; *Ulysses*
technology: and aesthetics, 2–6, 17–18, 28–34, 36–40, 189–196; and avant-garde, 2, 197n. 3; as constitutive of *Remembrance*, 94, 189–94; and death, 12–13, 15, 180–83, 189; definition of, 5–6, 46; effects on human sensorium, 1–2, 7–10, 15–19, 51–53, 189–96; as extension of body, 36, 52; and gender, 11, 36; Kittler's theory of, 44–46; Latour's theory of, 41–43; in *Magic Mountain*, 87–90, 189–94; and mass culture, 7–8, 31, 35; and modernism, 10–11, 17–18, 21–23, 38–39, 45–46, 54, 189–96; naturalization of, in *Ulysses*, 151, 191; Proust's theory of, 11–17; of the self in *Magic Mountain*, 211n. 39; as socially embedded, 40, 87–88. *See also* technologies of perception; technologies of speed; *and entries for individual technologies*
telephony, 2, 5; and death, 12–17, 180–83, 232nn. 55, 56; effects on aural experience, 12–17, 149, 156–57; effects on visual experience, 12–17; in Eliot, 47; in Rathenau, 181–82, 232n. 56; in *Remembrance*, 11–17, 157, 181, 190; in *Ulysses*, 23, 180–82; as network in Latour, 42; in Woolf, 157
temporality: in Bergson, 51; and chronophotography, 100; and concept of modernism, 28, 32–33; in *Magic Mountain*, 59–60; of modernity, 25, 43–44, 51; and photography, 196; private and public time, 9, 44, 92–93; in *Remembrance*, 51, 92–93, 120–21, 123, 183; in *Ulysses*, 182–84. *See also* experience; memory
thaumatrope, 63. *See also* optical devices
Theall, Donald F., 226n. 10
Thomet, Ulrich, 208n. 12
Tichi, Cecilia, 198n. 8
time. *See* temporality
time-motion studies. *See* chronophotography
Tolstoy, Leo, 128, 163
Torgovnick, Marianna, 202n. 8
total work of art. *See* Gesamtkunstwerk
touch. *See* tactility
tram: in Joyce, 191